ξ Zeta

Focus: Decimals and Percents

Instruction Manual

By Steven P. Demme

1-888-854-MATH (6284)
www.MathUSee.com

# Math·U·See

1-888-854-MATH (6284)
www.MathUSee.com
Copyright © 2009 by Steven P. Demme

All rights reserved. No part of this book may be reproduced, stored in a retrieval system, or transmitted in any form by any means—electronic, mechanical, photocopying, recording, or otherwise.

In other words, thou shalt not steal.

Printed in the United States of America

ξ Zeta

SCOPE AND SEQUENCE
HOW TO USE MATH-U-SEE
SUPPORT AND RESOURCES

LESSON 1	Exponents; Word Problem Tips
LESSON 2	Place Value
LESSON 3	Decimal Numbers with Expanded Notation
LESSON 4	Add Decimal Numbers
LESSON 5	Subtract Decimal Numbers
LESSON 6	Metric System Origin–Meter, Liter, Gram
LESSON 7	Metric System–Latin Prefixes
LESSON 8	Metric System Conversion–Part 1
LESSON 9	Multiply by 1/10 or .1
LESSON 10	Multiply Decimals by 1/100 or .01
LESSON 11	Finding a Percent of a Number
LESSON 12	Finding a Percent > 100%; Word Problems
LESSON 13	Reading Percents in a Pie Graph
LESSON 14	Multiply All Decimals
LESSON 15	Metric System Conversions – Part 2
LESSON 16	Computing Area and Circumference
LESSON 17	Dividing a Decimal by a Whole Number
LESSON 18	Dividing a Whole Number by a Decimal
LESSON 19	Solving for an Unknown
LESSON 20	Dividing Decimal by a Decimal
LESSON 21	Decimal Remainders
LESSON 22	More Solving for an Unknown
LESSON 23	Transform Any Fraction
LESSON 24	Decimals as Rational Numbers
LESSON 25	Mean, Median, and Mode
LESSON 26	Probability
LESSON 27	Points, Lines, Rays, and Line Segments
LESSON 28	Planes and Symbols
LESSON 29	Angles
LESSON 30	Types of Angles

STUDENT SOLUTIONS
TEST SOLUTIONS

GLOSSARY OF TERMS
MASTER INDEX FOR GENERAL MATH
ZETA INDEX

SCOPE & SEQUENCE

Math-U-See is a complete and comprehensive K-12 math curriculum. While each book focuses on a specific theme, Math-U-See continuously reviews and integrates topics and concepts presented in previous levels.

Primer

α Alpha | Focus: Single-Digit Addition and Subtraction

β Beta | Focus: Multiple-Digit Addition and Subtraction

γ Gamma | Focus: Multiplication

δ Delta | Focus: Division

ε Epsilon | Focus: Fractions

ζ Zeta | Focus: Decimals and Percents

Pre-Algebra

Algebra 1

Stewardship*

Geometry

Algebra 2

Pre Calculus with Trigonometry

*Stewardship is a biblical approach to personal finance. The requisite knowledge for this curriculum is a mastery of the four basic operations, as well as fractions, decimals, and percents. In the Math-U-See sequence these topics are thoroughly covered in Alpha through Zeta. We also recommend Pre-Algebra and Algebra 1 since over half of the lessons require some knowledge of algebra. Stewardship may be studied as a one-year math course or in conjunction with any of the secondary math levels.

HOW TO USE

Five Minutes for Success

Welcome to *Zeta*. I believe you will have a positive experience with the unique Math-U-See approach to teaching math. These first few pages explain the essence of this methodology which has worked for thousands of students and teachers. I hope you will take five minutes and read through these steps carefully.

I am assuming your student has a thorough grasp of the four basic operations (addition, subtraction, multiplication, and division), along with a mastery of fractions.

If you are using the program properly and still need additional help, you may contact your authorized representative, or visit Math-U-See online at http://www.mathusee.com/support.html

— S. Demme

The Goal of Math-U-See

The underlying assumption or premise of Math-U-See is that the reason we study math is to apply math in everyday situations. Our goal is to help produce confident problem solvers who enjoy the study of math. These are students who learn their math facts, rules, and formulas and are able to use this knowledge to solve word problems and real life applications. Therefore, the study of math is much more than simply committing to memory a list of facts. It includes memorization, but it also encompasses learning the underlying concepts of math that are critical to successful problem solving.

More than Memorization

Many people confuse memorization with understanding. Once while I was teaching seven junior high students, I asked how many pieces they would each receive if there were fourteen pieces. The students' response was, "What do we do: add, subtract, multiply, or divide?" Knowing how to divide is important, understanding when to divide is equally important.

THE SUGGESTED 4-STEP MATH-U-SEE APPROACH

In order to train students to be confident problem solvers, here are the four steps that I suggest you use to get the most from the Math-U-See curriculum.

Step 1. Prepare for the Lesson
Step 2. Present the New Topic
Step 3. Practice for Mastery
Step 4. Progression after Mastery

Step 1. Prepare for the Lesson.

Watch the DVD to learn the new concept and see how to demonstrate this concept with the manipulatives when applicable. Study the written explanations and examples in the instruction manual. Many students watch the DVD along with their instructor.

Step 2. Present the New Topic

Present the new concept to your student. Have the student watch the DVD with you, if you think it would be helpful. Older students may watch the DVD on their own.

- **a. Build:** Use the manipulatives to demonstrate the problems from the worksheet.

- **b. Write:** Record the step-by-step solutions on paper as you work them through with manipulatives.

- **c. Say:** Explain the *why* and *what* of math as you build and write.

Do as many problems as you feel are necessary until the student is comfortable with the new material. One of the joys of teaching is hearing a student say *"Now I get it!"* or *"Now I see it!"*

Step 3. Practice for Mastery.

Using the examples and the lesson practice problems from the student text, have the students practice the new concept until they understand it. It is one thing for students to watch someone else do a problem, it is quite another to do the same problem themselves. Do enough examples together until they can do them without assistance.

Do as many of the lesson practice pages as necessary (not all pages may be needed) until the students remember the new material and gain understanding. Give special attention to the word problems, which are designed to apply the concept being taught in the lesson.

Another resource is the Math-U-See web site which has online drill and downloadable worksheets for several of the lessons. Click on www.mathusee.com and select "Online Helps."

Step 4. Progression after Mastery.

Once mastery of the new concept is demonstrated, proceed to the systematic review pages for that lesson. Mastery can be demonstrated by having each student teach the new material back to you. The goal is not to fill in worksheets, but to be able to teach back what has been learned.

The systematic review worksheets review the new material as well as provide practice of the math concepts previously studied. Remediate missed problems as they arise to ensure continued mastery.

Proceed to the lesson tests. These were designed to be an assessment tool to help determine mastery, but they may also be used as extra worksheets. Your students will be ready for the next lesson only after demonstrating mastery of the new concept and continued mastery of concepts found in the systematic review worksheets.

Confucius was reputed to have said, "Tell me, I forget; Show me, I understand; Let me do it, I will remember." To which we add, **"Let me teach it and I will have achieved mastery!"**

Length of a Lesson

So how long should a lesson take? This will vary from student to student and from topic to topic. You may spend a day on a new topic, or you may spend several days. There are so many factors that influence this process that it is impossible to predict the length of time from one lesson to another. I have spent three days on a lesson and I have also invested three weeks in a lesson. This occurred in the same book with the same student. If you move from lesson to lesson too quickly without the student demonstrating mastery, he will become overwhelmed and discouraged as he is exposed to more new material without having learned the previous topics. But if you move too slowly, your student may become bored and lose interest in math. I believe that as you regularly spend time working along with your student, you will sense when is the right time to take the lesson test and progress through the book.

By following the four steps outlined above, you will have a much greater opportunity to succeed. Math must be taught sequentially, as it builds line upon line and precept upon precept on previously learned material. I hope you will try this methodology and move at your student's pace. As you do, I think you will be helping to create a confident problem solver who enjoys the study of math.

ONGOING SUPPORT
AND ADDITIONAL RESOURCES

Welcome to the Math-U-See Family!

Now that you have invested in your children's education, I would like to tell you about the resources that are available to you. Allow me to introduce you to your regional representative, our ever improving website, the Math-U-See blog, our new free e-mail newsletter, the online Forum, and the Users Group.

Most of our regional **Representatives** have been with us for over 10 years. What makes them unique is their desire to serve and their expertise. They have all used Math-U-See and are able to answer most of your questions, place your student(s) in the appropriate level, and provide knowledgeable support throughout the school year. They are wonderful!

Come to your local curriculum fair where you can meet your rep face-to-face, see the latest products, attend a workshop, meet other MUS users at the booth, and be refreshed. We are at most curriculum fairs and events. To find the fair nearest you, click on "Events Calendar" under "News."

The **Website**, at www.mathusee.com, is continually being updated and improved. It has many excellent tools to enhance your teaching and provide more practice for your student(s).

ONLINE DRILL

Let your students review their math facts online. Just enter the facts you want to learn and start drilling. This is a great way to commit those facts to memory.

WORKSHEET GENERATOR

Create custom worksheets to print out and use with your students. It's easy to use and gives you the flexibility to focus on a specific lesson. Best of all — it's free!

Math-U-See Blog

Interesting insights and up-to-date information appear regularly on the Math-U-See Blog. The blog features updates, rep highlights, fun pictures, and stories from other users. Visit us and get the latest scoop on what is happening.

Email Newsletter

For the latest news and practical teaching tips, sign up online for the free Math-U-See e-mail newsletter. Each month you will receive an e-mail with a teaching tip from Steve as well as the latest news from the website. It's short, beneficial, and fun. Sign up today!

The Math-U-See Forum and the Users Group put the combined wisdom of several thousand of your peers with years of teaching experience at your disposal.

Online Forum

Have a question, a great idea, or just want to chitchat with other Math-U-See users? Go to the online forum. You can also use the forum to post a specific math question if you are having difficulty in a certain lesson. Head on over to the forum and join in the discussion.

Yahoo Users Group

The MUS-users group was started in 1998 for lovers and users of the Math-U-See program. It was founded by two home-educating mothers and users of Math-U-See. The backbone of information and support is provided by several thousand fellow MUS users.

For Specific Math Help

When you have watched the DVD instruction and read the instruction manual and still have a question, we are here to help. Call your local rep, click the support link and e-mail us here at the home office, or post your question on the forum. Our trained staff have used Math-U-See themselves and are available to answer a question or walk you through a specific lesson.

Feedback

Send us an e-mail by clicking the feedback link. We are here to serve you and help you teach math. Ask a question, leave a comment, or tell us how you and your student are doing with Math-U-See.

Our hope and prayer is that you and your students will be equipped to have a successful experience with math!

Blessings,

Steve

Steve Demme

LESSON 1

Exponents; Word Problem Tips

Multiplication is fast adding of the same number. To take this a step further, fast multiplying of the same number is raising to a power, or *exponents*. You can think of exponents in several ways. Picture a square that is 10 over and 10 up. It is 10 two ways (over and up), it is also a square or "ten squared." By definition, the exponent—in this case 2—stands for how many times 10 is multiplied by itself. The number 10 is used as a factor twice.

Another way to state it is "ten to the two power" or "ten to the power of two." Any number may be expressed as a square, such as "two squared" or "three squared." Look at the figure 1 for some examples. Notice the many ways to express the same reality.

Figure 1

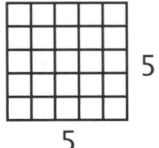 5
5

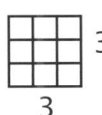 3
3

5 over and 5 up
5 used as a factor 2 times
5 to the second power
5 squared
5 to the 2 power
5 raised to the power of 2

$5^2 = 5 \cdot 5 = 25$

3 over and 3 up
3 used as a factor 2 times
3 to the second power
3 squared
3 to the 2 power
3 raised to the power of 2

$3^2 = 3 \cdot 3 = 9$

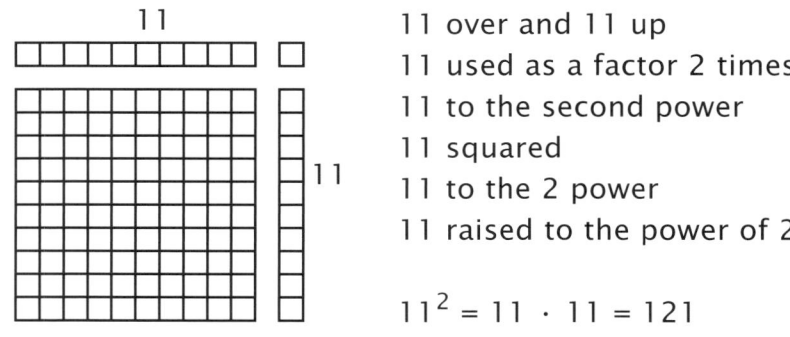

11 over and 11 up
11 used as a factor 2 times
11 to the second power
11 squared
11 to the 2 power
11 raised to the power of 2

$11^2 = 11 \cdot 11 = 121$

While it is difficult to show larger exponents with the manipulatives, you can have numbers raised to other powers besides two. For example $2 \times 2 \times 2 = 2^3$, which is 8. Another example is $3 \bullet 3 \bullet 3 \bullet 3 = 3^4$. It helps to say, "Three used as a factor four times is the same as 3^4, which is 81."

Example 1 $7^2 = 7 \cdot 7 = 49$

Example 2 $10 \times 10 \times 10 \times 10 \times 10 = 10^5 = 100,000$

When a number is raised to a power of three, such as 4^3, it is often read as "four to the third power," or "four cubed." You can make a figure that is a cube with dimensions 4 by 4 by 4. See figure 2.

Figure 2

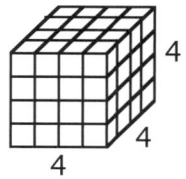

4 over and 4 up and 4 deep
4 used as a factor 3 times
4 to the third power
4 cubed
4 to the 3 power
4 raised to the power of 3

$4^3 = 4 \cdot 4 \cdot 4 = 64$

WORD PROBLEM TIPS

Parents often find it challenging to teach children how to solve word problems. Here are some suggestions for helping your student learn this important skill.

The first step is to realize that word problems require both reading and math comprehension. Don't expect a child to be able to solve a word problem if he does not thoroughly understand the math concepts involved. On the other hand, a student may have a math skill level that is stronger than his or her reading-comprehension skills. Below are a number of strategies to improve comprehension skills in the context of story problems. You should decide which ones work best for you and your child.

Strategies for word problems

1. Ignore numbers at first and read the story. It may help some students to read the question aloud. Every word problem tells a story. Before deciding what math operation is required, let the student retell the story in his own words. Who is involved? Are they receiving gifts, losing something, or dividing a treat?

2. Relate the story to real life, perhaps by using names of family members. For some students, this makes the problem more interesting and relevant.

3. Build, draw, or act out the story. Use the blocks or actual objects when practical. Especially in the lower levels, you may require the student to use the blocks for word problems even when the facts have been learned. Don't be afraid to use a little drama as well. The purpose is to make it as real and meaningful as possible.

4. Look for the common language used in a particular kind of problem. Pay close attention to the word problems on the lesson practice pages, as they model different kinds of language that may be used for the new concept just studied. For example, "altogether" indicates addition. These "key words" can be useful clues, but should not be a substitute for understanding.

5. Look for practical applications that use the concept and ask questions in that context.

6. Have the student invent word problems to illustrate number problems from the lesson.

Cautions:

1. Unneeded information may be included in the problem. For example, we may be told that Suzie is eight years old, but the eight is irrelevant when adding up the number of gifts she received.

2. Some problems may require more than one step to solve. Model these questions carefully.

3. There may be more than one way to solve some problems. Experience will help the student choose the easier or preferred method.

4. Estimation is a valuable tool for checking an answer. If an answer is unreasonable, it is possible that the wrong method was used to solve the problem.

LESSON 2

Place Value
with Expanded and Exponential Notation

In the base 10, or decimal system, all numbers are comprised of the digits 0 to 9 and place values: units, tens, hundreds, thousands, etc. There are several ways to take a number and break it down into these components. We will explore three of them. The first way, which should be review, is *place-value notation*. 156 is 100 + 50 + 6. The second way is *expanded notation*. 156 is 1x100 + 5x10 + 6x1. This shows the individual digits multiplied by each place value. The third way is *exponential notation*. Using what we've learned about exponents, we can express all the place values in expanded notation as 10 to a power.

In figure 1 on the next page, we show several of the place values, beginning with the units place and continuing through the ten thousands place. The first three are easily shown with the blocks as in the figure, but the thousands place poses a challenge. One way to build 1,000 is a cube that is 10 by 10 by 10. (See figure 2.) But if we follow the progression where the exponent indicates the dimension, as the second power for two dimensions and the third power as three dimensions, then we can't show 10,000, which would be the fourth dimension, because we can't draw or construct a fourth-dimensional figure. So we show each of the place values two-dimensionally. The number 1,000 is shown as 10 by 100. The number 10,000 is 100 by 100. If we had enough space, we could even show 100,000 as 100 by 1,000 and 1,000,000 as 1,000 by 1,000.

Notice the shapes as you move from right to left: square, rectangle, square, rectangle, square, . . . Place values with even exponents form squares, and the odd exponents produce rectangles.

Figure 1

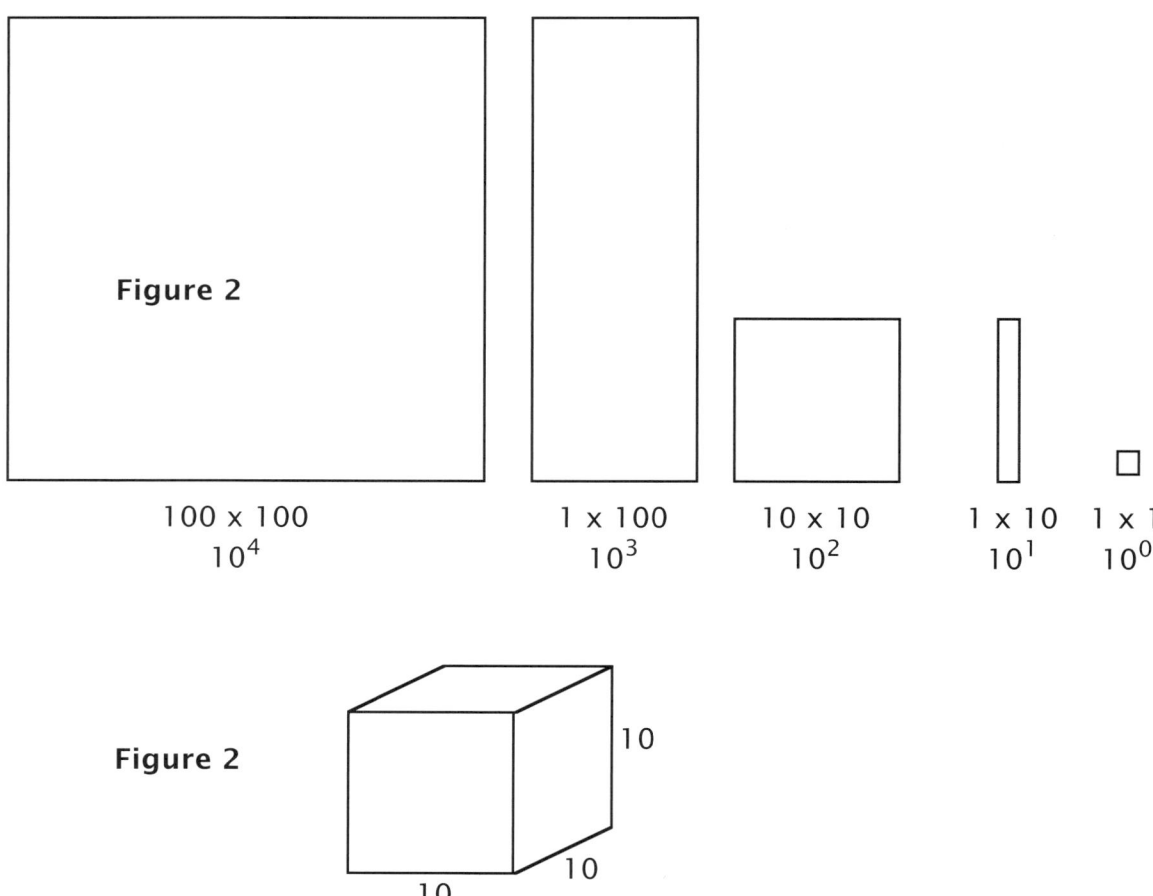

| 100 x 100 | 1 x 100 | 10 x 10 | 1 x 10 | 1 x 1 |
| 10^4 | 10^3 | 10^2 | 10^1 | 10^0 |

Figure 2

Notice how the place value corresponds to the powers of 10. We'll prove how $10^0 = 1$ in *Algebra 1*. See how the exponent corresponds to the number of zeros in the place value. When you multiply by 10, another zero is added. 10^3 is 1,000, which has three zeros, and 10^1 is 10, which has one zero. Using this knowledge, we'll show a number first in expanded notation, and then in exponential notation.

Figure 3

| 1,000,000 | 100,000 | 10,000 | 1,000 | 100 | 10 | 1 |
| 10^6 | 10^5 | 10^4 | 10^3 | 10^2 | 10^1 | 10^0 |

Example 1

Express 245, first in expanded notation and then with exponential notation.

$$245 = 2 \times 100 + 4 \times 10 + 5 \times 1$$
$$245 = 2 \times 10^2 + 4 \times 10^1 + 5 \times 10^0$$

Example 2

Express 1,759, first in expanded notation and then with exponential notation.

$$1,759 = 1 \times 1,000 + 7 \times 100 + 5 \times 10 + 9 \times 1$$
$$1,759 = 1 \times 10^3 + 7 \times 10^2 + 5 \times 10^1 + 9 \times 10^0$$

Example 3

Express 803, first in expanded notation and then with exponential notation.

$$803 = 8 \times 100 + 3 \times 1$$
$$803 = 8 \times 10^2 + 3 \times 10^0$$

Example 4

Express $5 \times 1,000 + 6 \times 100 + 4 \times 10$ as a number.

$$5 \times 1,000 + 6 \times 100 + 4 \times 10 = 5,640$$

LESSON 3

Decimal Numbers with Expanded Notation

Decimal numbers are fractions written in the base 10 system. They are a blend of what we know of place value and of fractions. We will go back and forth between these two concepts. If you keep this in mind throughout our study of decimals, you will be in good shape. This lesson will go back and forth between fractions and place value to show where decimals originated and how they relate to all we have learned up to this point. We are interested not only in learning how to add, subtract, multiply, and divide decimals, but also in understanding the concepts that shape our understanding of this subject.

Some teachers refer to decimals as *decimal fractions.* Perhaps this is because decimals are fractions written in the base 10, or decimal, system. In figure 1, a fraction is changed to an equivalent fraction with a denominator of 10. After constructing 2/5 with the overlays, place the clear overlay with 10 spaces to make tenths on top of 2/5. The result is 4/10 or four-tenths.

Figure 1

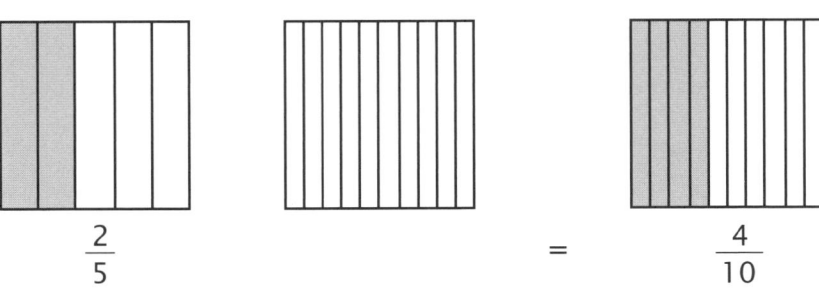

In expanded notation, 34 is 3x10 + 4x1. The digits are 3 and 4, and the values are ten and one. A fraction may also be written in expanded notation. The fraction 2/5 can be written as 2 x 1/5. The digit 2 tells how many, and the value 1/5 tells what kind. But in the decimal system, there is no fifths place. In fact, up to this point, we have not dealt with any place value less than one (units). So we need to develop some smaller place values.

First we'll do a little review. Recall that among the place values, as you move from right to left to increase the values, you multiply by a factor of 10. As you decrease and move from the larger to the smaller values and from left to right, you divide by a factor of 10. See figure 2.

Figure 2

$$\overline{100{,}000} \;\swarrow\; \overline{10{,}000} \;\swarrow\; \overline{1{,}000} \;\swarrow\; \overline{100} \;\swarrow\; \overline{10} \;\swarrow\; \overline{1}$$
$$10 \cdot 10{,}000 \quad 10 \cdot 1{,}000 \quad 10 \cdot 100 \quad 10 \cdot 10 \quad 10 \cdot 1$$

$$\overline{100{,}000} \;\searrow\; \overline{10{,}000} \;\searrow\; \overline{1{,}000} \;\searrow\; \overline{100} \;\searrow\; \overline{10} \;\searrow\; \overline{1}$$
$$100{,}000 \div 10 \quad 10{,}000 \div 10 \quad 1{,}000 \div 10 \quad 100 \div 10 \quad 10 \div 10$$

Since we are looking for values less than one, or fractional values, we need to continue this pattern of decreasing by or dividing by a factor of 10. From studying fractions, we know that dividing by 10 is the same as multiplying by 1/10, since the symbol "/10" means "divided by 10." Picking up where we left off in figure 2, we'll divide 1 by 10 or multiply 1 by 1/10. See figure 3.

Figure 3

$$\overline{1} \;\searrow\; \overline{\tfrac{1}{10}} \;\searrow\; \overline{\tfrac{1}{100}} \;\searrow\; \overline{\tfrac{1}{1{,}000}}$$

$$1 \div 10 \quad\quad 1/10 \div 10 \quad\quad 1/100 \div 10$$
$$1 \times 1/10 \quad\quad 1/10 \times 1/10 \quad\quad 1/100 \times 1/10$$

All we need now is a symbol to separate the whole-number place values from the fractional place values. As in money, the symbol is a decimal point. Between the units place and the tenths place, we put a decimal point. See figure 4 for our new fraction values with the old number values from hundreds to hundredths. Notice that the units place, and not the decimal point, is the pivotal place.

Figure 4

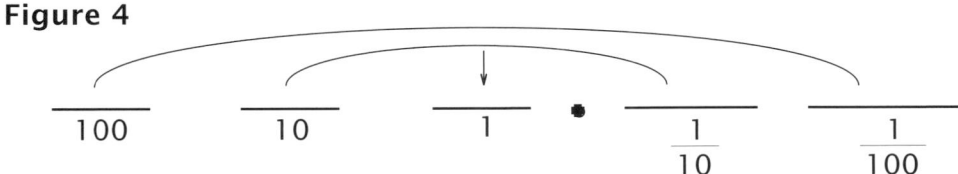

Since we have values that are less than one unit, refer back to figure 2, where we changed 2/5 to 4/10. We did this because there is not a 1/5 place in the decimal system. However, there is a 1/10 place. In expanded notation, 4/10 is written as 4 x 1/10. The fraction 4/10 may also be written as the decimal .4 (figure 5). It can be expressed as 4 in the tenths place, or 4 tenths, or .4.

Figure 5

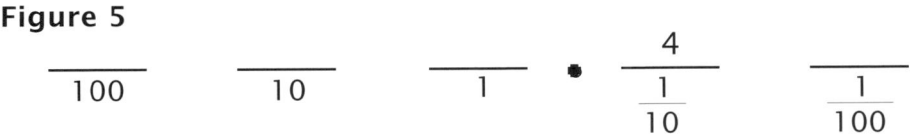

Money is a pure decimal function. If one unit is one dollar, 1/10 of a dollar is one dime, and 1/100 of a dollar (or 1/10 of a dime) is one penny. Two-fifths of a dollar is the same as 4/10 of a dollar or four dimes, as shown in figure 5. We will explore more about money on the worksheets and in another lesson.

Study the examples below to add to your understanding of decimals.

Example 1
Express 34.762 with expanded notation.

$$34.762 = 3 \times 10 + 4 \times 1 + 7 \times \frac{1}{10} + 6 \times \frac{1}{100} + 2 \times \frac{1}{1,000}$$

Example 2
Express 590.184 with expanded notation.

$$590.184 = 5 \times 100 + 9 \times 10 + 1 \times \frac{1}{10} + 8 \times \frac{1}{100} + 4 \times \frac{1}{1,000}$$

Because of what we know of exponents, we can rewrite examples 1 and 2 in what is called exponential notation. We will use 10^0 to represent the units place. (We'll learn in *Algebra 1* why $10^0 = 1$.)

Example 1 (in exponential notation)

$$34.762 = 3 \times 10^1 + 4 \times 10^0 + 7 \times \frac{1}{10^1} + 6 \times \frac{1}{10^2} + 2 \times \frac{1}{10^3}$$

Example 2 (in exponential notation)

$$590.184 = 5 \times 10^2 + 9 \times 10^1 + 1 \times \frac{1}{10^1} + 8 \times \frac{1}{10^2} + 4 \times \frac{1}{10^3}$$

LESSON 4

Add Decimal Numbers

In this lesson, you get to meet the pieces that represent the decimals. Turn a red hundred square upside down so the hollow side is showing, and snap the flat green piece (from the algebra/decimal inserts) into the back. Then turn over several blue 10 bars and snap the flat blue pieces (also from the inserts) into their backs. Then take out the little one-half inch red cubes.

The large green square represents one unit. We've increased the size of the unit from the little green cube to this larger size, just as we did when learning fractions. Since the large green square represents one, what do you think the flat blue bars represent? It takes ten of them to make one, so they are each 1/10 or .1. The red cubes represent 1/100 or .01.

In figure 1, we represent 1.56 or 1 x 1 + 5 x 1/10 + 6 x 1/100 with the decimal inserts.

Figure 1

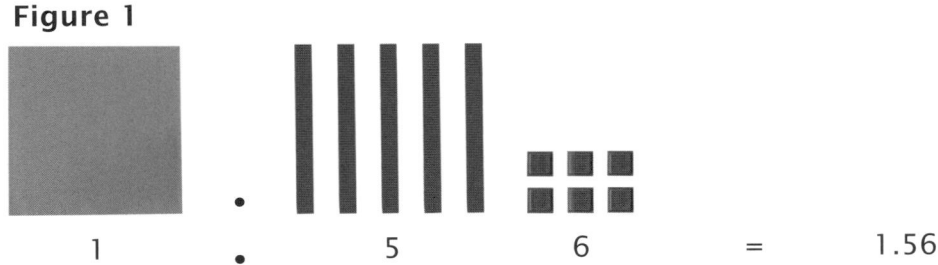

As we've said before, money is a pure decimal function. If figure 1 represents money with the green unit as one dollar, then 1/10 of a dollar is one dime and is represented by the blue 1/10 bars. As shown with the red cubes, 1/100 of a dollar or 1/10 of a dime is one penny.

Figure 2

This will help us in adding and subtracting decimals. The key to understanding this is the old adage, "to compare or combine, you must be the same kind." You can only add dollars to dollars and dimes to dimes and pennies to pennies. So also in decimals, you can only add units to units and tenths to tenths and hundredths to hundredths. The easiest way to distinguish the values and make sure you are combining like values is by writing numbers vertically, so the decimal point in one number is directly above (or below) the decimal point in the other number. Lining up these points ensures you that your place values are also lined up. You may only add or subtract two numbers if they have the same value.

When using the inserts, it is clear that you can only add the green to the green, the blue to the blue, etc. But when we don't have the inserts for larger numbers, always line up the decimal points. The same skills are used for adding decimals and money as for adding any number. Remember that decimals are pure base 10. You've just learned some new kinds of decimal values.

Example 1
Add 1.56 + 1.23

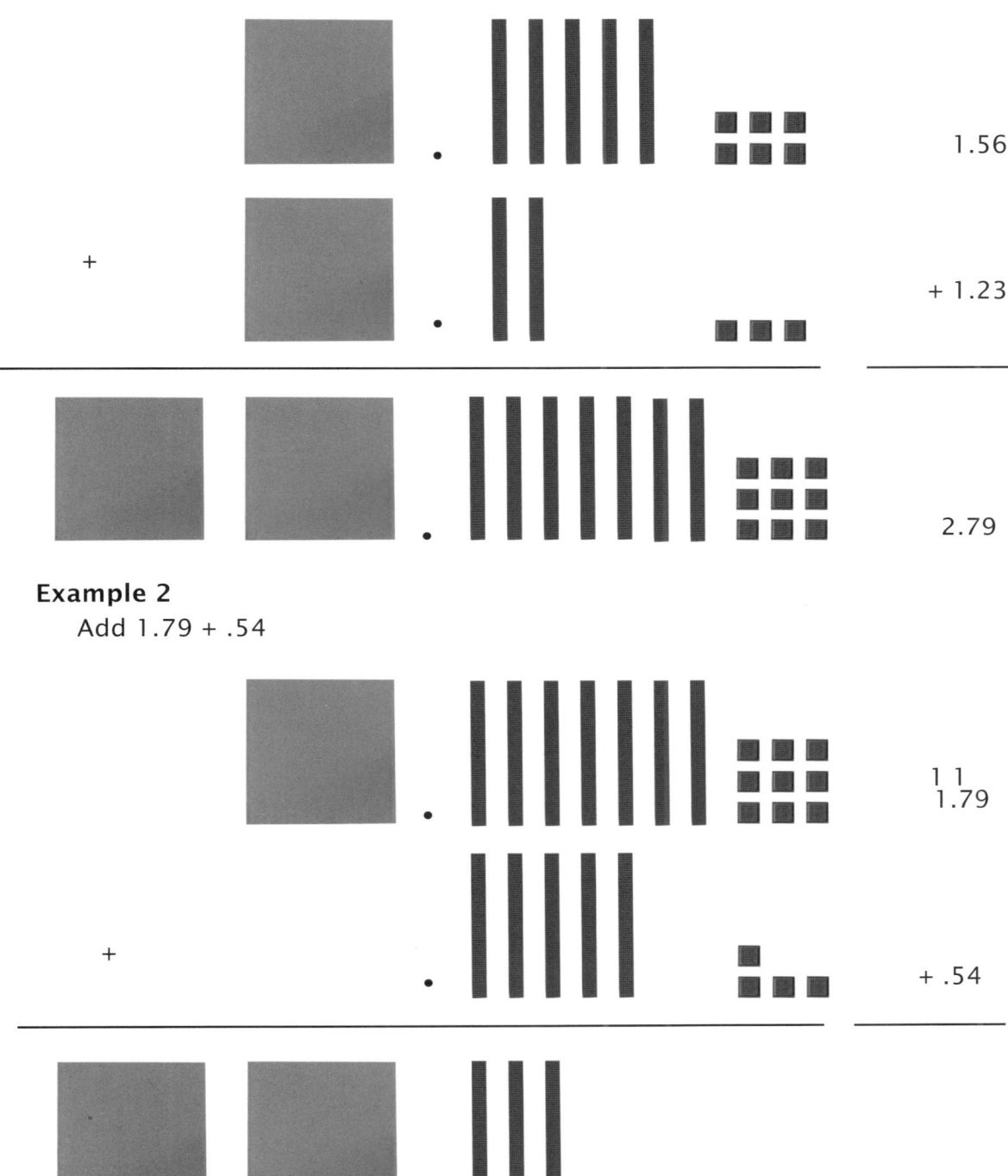

Example 2
Add 1.79 + .54

LESSON 5

Subtract Decimal Numbers

When subtracting decimals, the same rules of regrouping apply once you have lined up the place values. It helps to think of money when adding and subtracting decimals. In examples 1 and 2, subtraction problems are illustrated. There are no blocks or inserts to represent subtraction, but at this stage the student should be able to subtract accurately and readily.

Example 1 Subtract 1.76 - 1.42

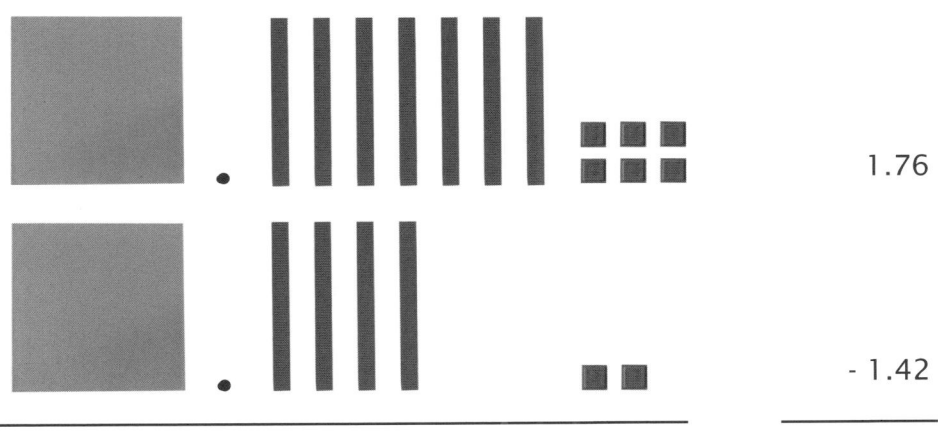

```
  1.76
- 1.42
------
   .34
```

Example 2   Subtract 2.29 - 1.75

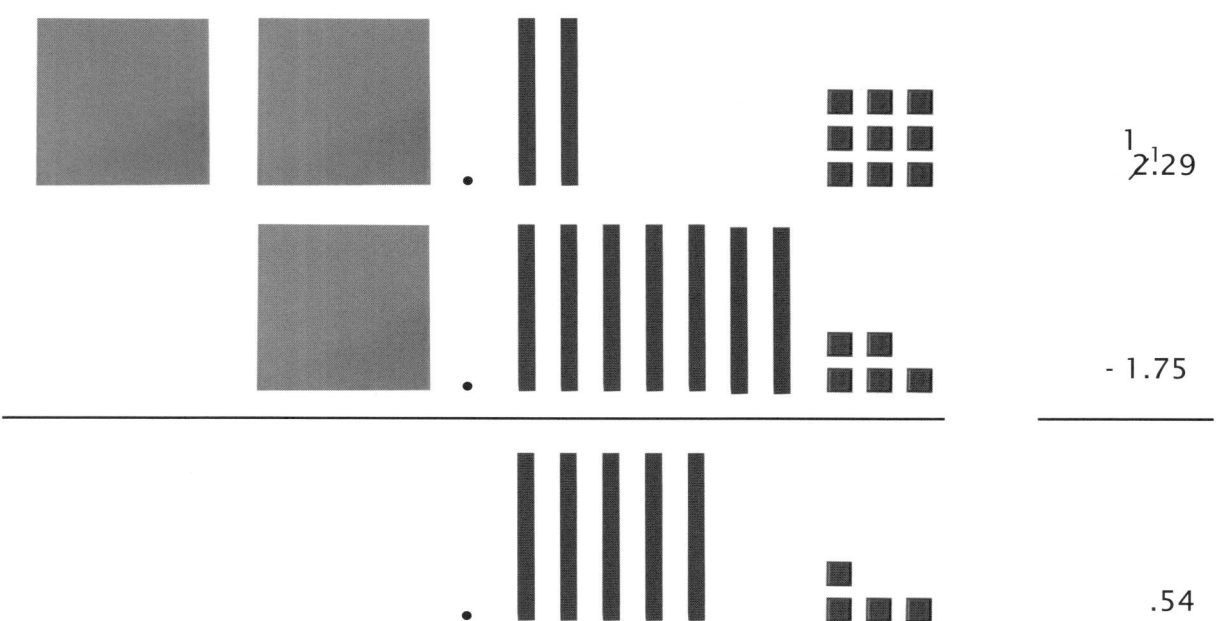

$$\begin{array}{r} \overset{1}{2}\overset{}{.}29 \\ -\,1.75 \\ \hline .54 \end{array}$$

LESSON 6

Metric System Origin—Meter, Liter, Gram
and Greek Prefixes; Multi-step Word Problems

Metric comes from "metre," meaning measure. In the 1790s, a group of French scientists were commissioned (government job) to create another form of measurement. They began with the distance from the North Pole to the equator and divided it up into 10,000,000 equal pieces. Each piece was one *meter*, or metre, depending on how you spell it.

The key for understanding the interrelationships between the different types of measurement in the metric system is found in the *centimeter*, which is 1/100 of a meter. A *cubic centimeter*, or cc, is approximately the size of the inside of a green unit block. When this is filled with liquid, it represents a *milliliter*. If the liquid is water, the mass or weight of the water is one *gram*. Hence 1 cc (linear measure) = 1 ml (liquid measure) = 1 g (mass or weight). The key words for the three different types of measure are:

Linear Measure	Liquid Measure	Mass or Weight
meter - m	liter - L	gram - g

Once these words are recognized, the next step is to learn the different prefixes. Each of these prefixes corresponds to a place value in the decimal system. The larger prefixes, representing 10, 100, and 1,000, are of Greek derivation. They, like the word Greek, all have a hard "k" sound.

kilo - k	hecto - h	deka (or deca) - dk (or da)
1,000	100	10

When you see kilometer, you know you have 1,000 meters. Kilogram means 1,000 grams and kiloliter represents 1,000 liters. Each term has an abbreviation as shown above. When you put the unit of measure together with the appropriate prefix, you have something similar to a number. The prefix tells you how many, and the measure tells you what kind. For example, in hg the "h" means hecto or 100, and the "g" stands for grams. So hg means 100 grams. In the same way, dkl means 10 liters. The "dk" is for deka or 10, and the "L" for liters. The abbreviation for liter is written with an uppercase L to avoid confusion. You may also see it written in cursive. Deka with the dk abbreviation is used in the U.S., although you may also see "deca" with the abbreviation "da." Deca and da are the accepted international forms.

Example 1

Fill in the spaces above and below the lines with the appropriate word, number and abbreviation.

1. | kilogram (kg) | _____ | _____ | _____ |
   |---|---|---|---|
   | 1,000 g | 100 g | 10 g | 1 g |

2. | _____ | _____ | dekaliter (dk) | _____ |
   |---|---|---|---|
   | | 100 L | | 1 L |

3. | _____ | _____ | _____ | meter (m) |
   |---|---|---|---|
   | 1,000 m | | | |

Example 1 solution

1. | kilogram (kg) | hectogram (hg) | dekagram | gram (g) |
   |---|---|---|---|
   | 1,000 g | 100 g | 10 g | 1 g |

2. | kiloliter (kl) | hectoliter (hl) | dekaliter (dk) | liter (L) |
   |---|---|---|---|
   | 1,000 L | 100 L | 10 L | 1 L |

3. | kilometer (km) | hectometer (hm) | dekameter | meter (m) |
   |---|---|---|---|
   | 1,000 m | 100 m | 10 m | 1 m |

MULTI-STEP WORD PROBLEMS

The student text includes some fairly simple two-step word problems. Some students may be ready for more challenging problems. Here are a few to try, along with some tips for solving this kind of problem. You may want to read and discuss these with your student as you work out the solutions together. The purpose is to stretch, not to frustrate. If you do not think the student is ready, you may want to come back to these later.

There are more multi-step word problems on the next page and in lessons 12, 18, and 24 of this instruction manual. The answers are at the end of the solutions at the back of this book.

1. Jill bought items costing $3.45, $1.99, $6.59, and $12.98. She used a coupon worth $2.50. If Jill had $50.00 when she went into the store, how much did she have when she left?

Although the problem asks only one question, there are other questions that must be answered first. The key to solving the problem is determining what the unstated questions are. Since the final question is asking for the leftover money, the unstated questions are: "What is the total of Jill's purchases?" and "What was the total bill after using the coupon?"

You might make a list of questions like this:

1. Total of purchases?

2. Total bill after subtracting coupon?

3. Leftover money?

2. Luke and Seth started out to visit Uncle Arnie. After driving 50 miles, they saw a restaurant, and Luke wanted to stop for lunch. Seth wanted to look for something better, so they drove on for eight miles before giving up and going back to the restaurant. After eating, they traveled on for 26 more miles from the restaurant. Seth saw a sign for a classic car museum, which they decided to visit. The museum was six miles from their route. After returning to the main road, they drove for another 40 miles and arrived at Uncle Arnie's house. How many miles is it from Luke and Seth's house to Uncle Arnie's house? How many miles did they drive on the way there?

The key to solving this is a careful drawing. It does not have to be to scale, but it should include all the parts of the journey.

3. Last week we got 3.5 inches of snow. Six-tenths of an inch melted before another storm added 8.3 inches. Since then we have lost 4.2 inches to melting or evaporation. How may inches of snow are left on the ground?

LESSON 7

Metric System–Latin Prefixes

Latin supplies the three prefixes that describe the smaller units of measure. Even though *milli* in Latin means 1,000, it is used in the metric system to represent 1/1000 (one thousandth). *Centi* is used for 1/100 (one hundredth) and *deci* represents 1/10 (one tenth). Their abbreviations are "d" for deci, "c" for centi, and "m" for milli. Notice that all three have an "i" in them. This may help you avoid confusing them with the prefixes for the larger units. They are also the small prefixes and have an "s" sound in the first two prefixes, like the "s" in small.

deci - d	centi - c	milli - m
1/10	1/100	1/1,000

Example 1

Fill in the spaces above and below the lines with the appropriate word, number and abbreviation.

1. gram (g) / 1 g , ___ / 1/10 g , ___ / ___ , ___ / ___

2. ___ / ___ , ___ / 100 L , ___ / 1/100 L , milliliter (ml) / ___

3. ___ / ___ , ___ / ___ , centimeter (cm) / ___ , ___ / 1/1,000 m

Example 1 solution

1.
gram (g)	decigram (dg)	centigram (cg)	milligram (mg)
1 g	1/10 g	1/100 g	1/1,000 g

2.
liter (L)	deciliter (dl)	centiliter (cl)	milliliter (ml)
1 L	1/10 L	1/100 L	1/1,000 L

3.
meter (m)	decimeter (dm)	centimeter (cm)	millimeter (mm)
1 m	1/10 m	1/100 m	1/1,000 m

Somtimes, the uppercase "L" is used in combination with other letters. For example, ml may be written as mL.

You may wonder how these correspond to units of measure in imperial measurement. There are a few that I think are worth remembering:

One meter is a little over three feet (ft), or one yard (yd).
One meter is about 39.37 inches (in), while one yard (three feet), is 36 inches.

One centimeter is a little less than one-half (.5) of an inch. It is close to .4 of an inch. 2.5 cm or 2 1/2 cm ≈ one inch.

One liter is approximately 1.06 quarts (qt).

One ounce (oz) is approximately 28 grams.
One gram is about the weight of a small paper clip.

One kilogram is approximately 2.2 pounds (lb).

One kilometer is a little over one-half mile (mi), or approximately .6 of a mile.

LESSON 8

Metric System Conversion–Part 1

When doing metric conversions, I like to put the smaller one on the right, as in place value where the larger units are to the left and the smaller to the right. Work them through step by step on the chart that is given.

kilo	hecto	deka	XXX	.	deci	centi	milli
1,000	100	10	1		$\frac{1}{10}$	$\frac{1}{100}$	$\frac{1}{1,000}$

Example 1

1 m = ___ cm

Put a finger on the centimeter place and ask, "How many centimeters in a decimeter?" The answer is 10. "How many decimeters in a meter?" "10." Every time you move from right to left on the chart (just as we do in the decimal system), you increase by a factor of 10. In this case we are moving two places, so we multiply by 10 twice. 10 x 10 = 100, so 1 m = 100 cm.

Example 2

1 km = ___ m

Put a finger on the meter place and ask, "How many meters in a dekameter?" "10." "How many dekameters in a hectometer?" "10." "How many hectometers in a kilometer?" "10." Every time you move from right to left on the chart, you increase by a factor of 10. In this case, we are moving three places, so we multiply by 10 three times. 10 x 10 x 10 = 1,000, so 1 km = 1,000 m.

Example 3
1 dkm = ___ mm

Put a finger on the millimeter place and ask, "How many millimeters in a centimeter?" "10." "How many centimeters in a decimeter?" "10." "How many decimeters in a meter?" "10." "How many meters in a dekameter?" "10." Every time you move from right to left on the chart you increase by a factor of 10. In this case, we are moving four places, so we multiply by 10 four times. 10 x 10 x 10 x10 = 10,000, so 1 dkm = 10,000 mm.

Up to this point, the student has been asked to change one kilogram to grams, or one of any unit to another unit. All this required was changing the values. One kilogram is 100 dekagrams or 1 kg = 100 dkg. Now he is being asked to change a number of kilograms to grams.

For example: 5 kg = ___ dkg. First find the change in value as before, and then multiply that times the number. Remember, when finding how many dekagrams in a kilogram, as in the example, you move two places. So we increase by a factor of 10 two times, or 10 x 10 = 100. Then we multiply by the 5, and 5 x 100 = 500.

Have the student think of money to get this concept. Five dollars = ___ pennies. Go through the above steps to see whether they help. There are 10 pennies in one dime, and 10 dimes in one dollar. So there are 10 x 10 = 100 pennies in one dollar. Since there are five dollars, there are 5 x 100 = 500 pennies in five dollars.

The most important aspect of this lesson is recognizing the place-value relationships. The numbers are secondary. So first find the proper place value, and then multiply the number.

Example 4
56 km = ___ dkm

1 km = 100 dkm

Multiply both sides by 56 and you will see that
56 km = 5,600 dkm.

LESSON 9

Multiply by 1/10 or .1

Multiplying decimals is based on what we already know of double-digit multiplication. Only our dimensions have changed. Recognizing the dimensions and area of the inserts is important, but understanding the difference between the the factors and the product is critical. If needed, review lesson 4, where we first introduced the decimal inserts, before going further.

The area, or product, of the green unit is one, and the dimensions, or factors, of the green unit square are one over and one up. The flat blue bar represents a product of 1/10 or .1, with an over factor of 1/10 and an up factor of one. Look at the figures below and think about this until you feel comfortable with the concept. Once you understand the blue bar, the red block will make more sense. It represents a product of 1/100 or .01, and the factors are 1/10 by 1/10, or .1 x .1. There is more explanation on the next page.

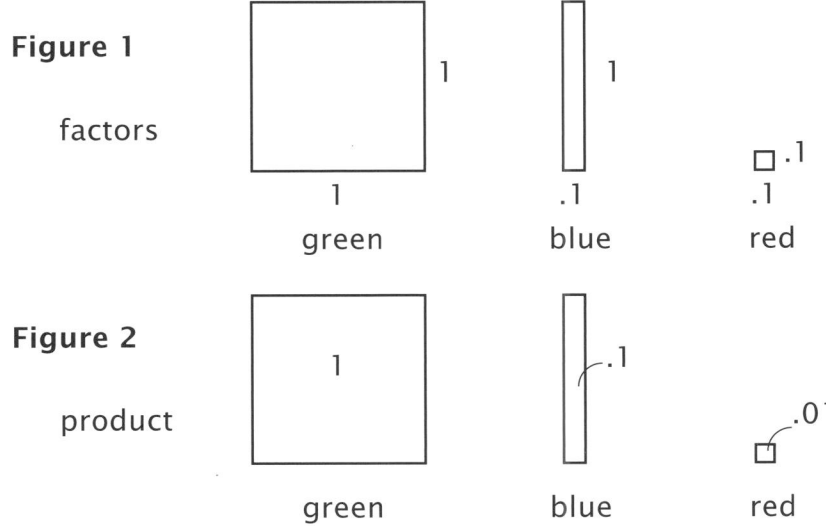

These figures look simple, but they are the key to understanding how and why to multiply decimals. Multiply the factors in figure 1 to find the product in figure 2. The unit block's factors are one and one, so one times one equals one. For the blue 1/10 bar, the factors are 1/10 and one, and 1/10 times one equals 1/10. With decimals this is .1 x 1 = .1.

The red 1/100 piece has factors 1/10 and 1/10. Thinking from what you know of fractions, you are taking 1/10 of 1/10. In the three-step process for taking a fraction of a fraction, you divide 1/10 into ten equal parts, and then take one of them. The result of 1/10 times 1/10 is 1/100. With decimals, it is .1 x .1 = .01. Money can help here. We know that 1/10 of a dime (which is 1/10 of a dollar) is a penny.

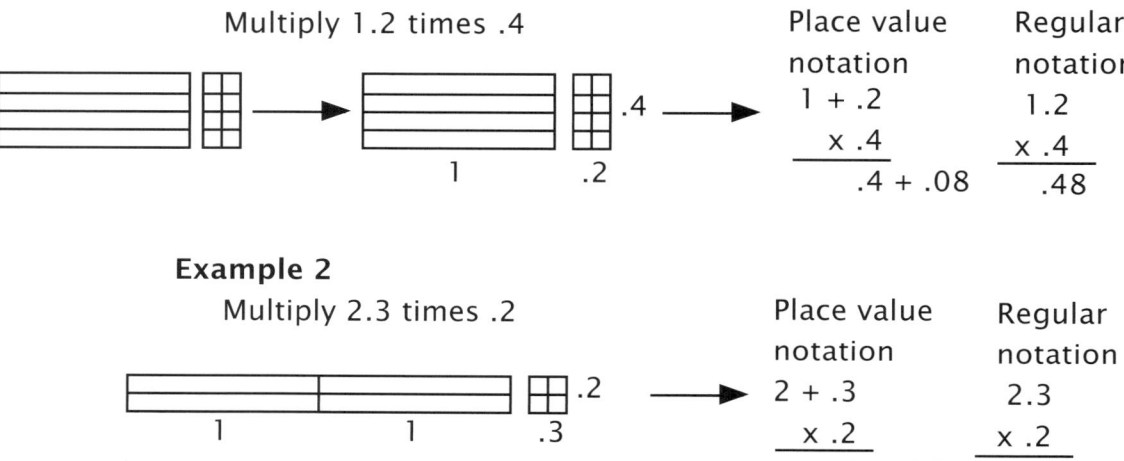

Example 3

Find the factors and area of this rectangle.

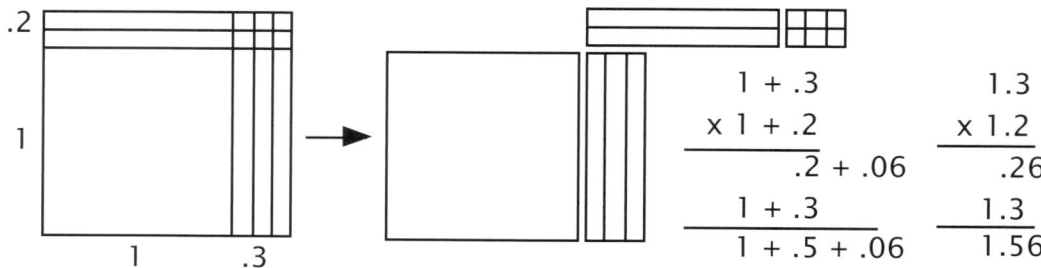

The factors are 1.3 by 1.2. The area is 1.56.

Notice that when building the rectangles, the up factor is on the line with the multiplication symbol, and the over factor is on the top line. Switching these factors will still produce the same answer, but it won't correspond with the picture.

I write this differently than most, so that the conceptual teaching comes first, and the rules follow the concept. I do this for two reasons.

First, most students expect a large answer when multiplying, as in 10 x 20 = 200. They usually see a product much larger than either factor. But don't forget, we are multiplying mixed numbers now, and our product is not going to be significantly larger than either factor.

Second, I want to reveal the reason for the rule normally employed when multiplying decimals, which is multiply, and then count the number of places to the right of the decimal point. Instead of moving the decimal point, we're going to do something "radical" and leave it stationary. Actually, moving decimal points is radical since they don't really move; it just appears that they do.

In the first step, .2 x .3 (or 2/10 x 3/10) shows that .2 x .3 is .06 (or 6/100). Placing this in the proper place, the 1/100s place, reveals two things. First, our product will be small, since we are starting two places to the right, and second, that is where the "multiply and then count" comes from. Tenths (one place over) times tenths (one place over) yields hundredths in our answer. One place over plus one place over equals two places over. Then .2 x 1 is .2, 1 x .3 is .3, and 1 x 1 is 1. When adding, always add the same values, so .06 + 0 is .06, .2 + .3 is .5, and 1 + 0 is 1. The answer is 1.56, as predicted.

Example 4

Find the factors and area of this rectangle.

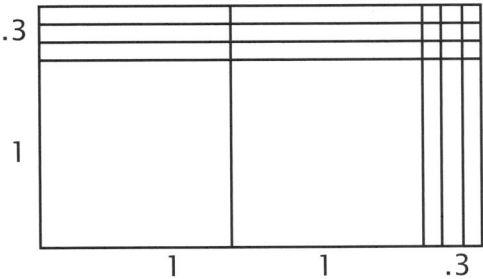

LESSON 10

Multiply Decimals by 1/100 or .01

Having mastered multiplying by tenths in the previous lesson, we are going to proceed to multiplying by hundredths. Emphasize thinking through each step and not merely memorizing. Even if the student chooses to multiply and then count as described later in the lesson, this thinking is still invaluable for estimating the final answer. We'll begin by taking 1/100 or .01 of several numbers.

Example 1
Find 1/100 of 1. Remember that "of" means "times."

$$\frac{1}{100} \text{ of 1 or } \frac{1}{100} \times \frac{1}{1} \text{ is } \frac{1}{100} \qquad .01 \times 1 = .01$$

Example 2
Find 1/100 of 100.

$$\frac{1}{100} \text{ of 100 or } \frac{1}{100} \times \frac{100}{1} \text{ is } \frac{100}{100} \text{ or 1} \qquad .01 \times 100 = 1$$

Example 3
Find 1/100 of 10.

$$\frac{1}{100} \text{ of 10 or } \frac{1}{100} \times \frac{10}{1} \text{ is } \frac{10}{100} \text{ or } \frac{1}{10} \qquad .01 \times 10 = .1$$

Example 4

Find 1/100 of 1/10.

$$\frac{1}{100} \text{ of } \frac{1}{10} \text{ or } \frac{1}{100} \times \frac{1}{10} \text{ is } \frac{1}{1,000} \qquad .01 \times .1 = .001$$

Notice that when you multiply by 1/100, it is the same as dividing by 100. When you divide by 100, your answer moves two places to the right because it is that much smaller. It appears that the decimal point "moves" two places to the left. These are two ways of describing the same phenomenon. Multiplying by 1/10 (which is .1 in decimal form) or dividing by 10 moves the answer one place to the right, and the decimal point appears to move one place to the left.

Example 5

```
    2.3              23           one place
  x .32            x 32           + two places
   .046             46
   .69      or      69
   .736            736
                  .736            three places from the right
```

We know how to multiply; the key is place value. You can think, "hundredths times tenths is thousandths" and begin in the thousandths place with the partial products. Or, you can multiply and solve the problem as if there were no decimals. Then count how many spaces to the right of the decimal point in both factors (one space and two spaces) and then count from the right and put the decimal point after the third space from the right.

This is the same as thinking hundredths times tenths is thousandths, multiplying as if there were no decimal points, and then placing the decimal point in the answer so the the last digit is in the thousandths place.

Example 6

```
    300.              300          zero places
  x  .02            x   2         + two places
    6.00    or      600
                   6.00            two places from the right
```

A hundredth times a hundred is 1.
2 x 3 = 6.

Example 7

```
    47.               47           zero places
  x .33             x 33          + two places
   1.41              141
  14.1      or       141
  15.51             1551
                   15.51           two places from the right
```

A hundredth times a unit is a hundredth.

Example 8

```
   4.32              432           two places
  x .65            x  65          + two places
   .2160            2160
  2.592             2592
  2.8080           28080

                   2.8080          four places from the right
```

A hundredth times a hundredth is a ten-thousandth.

LESSON 11

Finding a Percent of a Number

In a previous lesson we changed a fraction to a decimal by placing the tenth overlay on top of it. We can go a step further and place the other tenth overlay on top of the first one at a 90° angle. See figure 1.

Here we change 2/5 to 4/10 to 40/100. Notice that 4/10 = .4 and 40/100 = .40.

Figure 1

$$\frac{2}{5} = \frac{4}{10} = \frac{40}{100}$$

.4   .40

In figure 2, we take the 40/100 and show how that is transformed to a percent by taking the one and the two zeros from the number 100 and changing them into a percent sign. Now we see how to transform a fraction (2/5) to a decimal (.4) to a percent (40%).

Figure 2

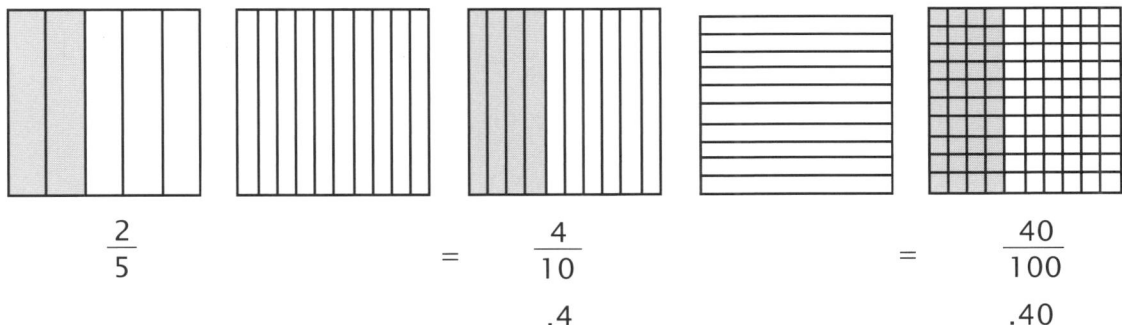

Percent means "per hundred". A percent is another way of writing hundredths. If you can change a fraction to hundredths, then you can easily change it to a percent. The converse is also true. If you have a percent, you can immediately restate it as hundredths. To take a percent of a number, simply change it to hundredths and multiply it by the number. See example 1.

Example 1
Find 25% of 36.
25% = .25 and .25 x 36 = 9. So 25% of 36 is 9.

Another way to solve the same problem would be to change 25% to a fraction, and then reduce it and multiply to find a fraction of a number. See example 2.

Example 2
Find 25% of 36.
$$25\% = \frac{25}{100} = \frac{1}{4} \qquad \frac{1}{4} \times 36 = 9$$

Both ways are legitimate. Some problems are easier to work using fractions and some using decimals.

Example 3
Find the tax on an order of onion rings that cost $1.50 if the tax rate is 8%.
8% = .08 and .08 x $1.50 = $.12

Example 4
Find the tip on a bill of $12.50 if you are tipping 16%.
16% = .16 and .16 x $12.50 = $2.00

Example 5
Burgers are 20% off on Tuesdays. How much is a $2.80 burger on Tuesday?
20% = .20 and .20 x $2.80 = $.56
$2.80 - .56 = $2.24

Not to confuse the issue, but there is another way to solve example 5. If you are going to be taking 20% off, then the final answer will be 100% minus 20% or 80% of the original amount. The total is 100%, or one. Then 80% is .80 and .80 x $2.80 = $2.24. Instead of finding 20% and multiplying, then subtracting, you are subtracting from 100% and then multiplying. Experiment by trying the same problem different ways to see which method is more comfortable for you.

There are several percentages that I have memorized that have proved helpful to me. I have listed them below in figure 3. We will be able to figure out some of the equivalents ourselves when we learn to divide decimals later in this book. Since one-fifth is 20%, then it follows two-fifths is double that, or 40%, three-fifths is 60%, and four-fifths is 80%.

Figure 3

$$\frac{1}{1} \times 1.00 = 100\%$$

$$\frac{1}{2} \times .50 = 50\%$$

$$\frac{1}{4} \times .25 = 25\%$$

$$\frac{3}{4} \times .75 = 75\%$$

$$\frac{1}{5} \times .20 = 20\%$$

LESSON 12

Finding a Percent > 100%; Word Problems

It is possible to have percents greater than 100%. We know 100% is the same as one. Consider the percents given in figure 1. Percents greater than one, or greater than 100%, are the same as numbers larger than one.

Figure 1

$$1 = \frac{100}{100} = 100\% \qquad 2 = \frac{200}{100} = 200\% \qquad 3 = \frac{300}{100} = 300\%$$

Figure 2 shows percents greater than one that are not whole numbers but mixed numbers.

Figure 2

$$1\frac{1}{2} = \frac{3}{2} = \frac{150}{100} = 150\% \text{ or}$$

$$1\frac{1}{2} = 1 + \frac{1}{2} = \frac{100}{100} + \frac{50}{100} = \frac{150}{100} = 50\%$$

$$2\frac{1}{4} = \frac{9}{3} = \frac{225}{100} = 225\% \text{ or}$$

$$2\frac{1}{4} = 2 + \frac{1}{4} = \frac{200}{100} + \frac{25}{100} = \frac{225}{100} = 150\%$$

In a problem that involves finding a tax, you can use the information on the previous page to save yourself some work.

If onion rings are $1.50 with a tax rate of 8%, then the tax is 8% or .08 x $1.50, which is $.12. Adding $1.50 + $.12 produces the the actual cost of buying the onion rings, which is $1.62.

Instead of multiplying by 8%, and then adding, you could also have multiplied by 1.08. You have to pay the total cost of the onion rings, which is 100% of the cost, plus the tax, which is 8% of the cost. So 108% is the cost of the rings and the tax inclusive. 100% is the same as 1.00 and 8% is the same as .08, so 108% is the same as 1.08. Multiply $1.50 times 1.08, and you get $1.62. Either method will work.

Example 1

Find the total cost for a meal priced at $12.50 if you are tipping 16% and the tax is 8%.

Total + tip + tax = 100% + 16% + 8% = 124%
124% x $12.50 = 1.24 x $12.50 = $15.50

WORD PROBLEMS

Here are a few more multi-step word problems to try. You may want to read and discuss these with your student as you work out the solutions together. Again, the purpose is to stretch, not to frustrate. If you do not think the student is ready, you may want to come back to these later.

The answers are at the end of the solutions at the back of this book.

1. Naomi was ordering yarn from a discount web site. If she ordered more than $50 worth of yarn labeled "discountable," she could take an extra 10% from the price of that yarn. Only some of the yarn she ordered was eligible for the discount. Here is what Naomi ordered:

    Two skeins at $2.50 per skein.   (not discountable)

    Eight skeins at $8.40 per skein.   (discountable)

    Seven skeins at $5.99 per skein.   (discountable)

    Tax is 5% and shipping is 8%. How much did Naomi pay for her yarn?

2. After Naomi received her order, she found that she could make a scarf from one skein of the yarn that cost $5.99 a skein. After taking into account the discount, taxes, and shipping for the yarn, how much would it cost to make five scarves?

3. Naomi made a scarf for her brother using the yarn that cost $2.50 per skein. He offered to pay Naomi for the yarn she used. If she used 1.5 skeins of yarn to make the scarf, what was the total cost of the yarn?

LESSON 13

Reading Percents in a Pie Graph

Circular *pie graphs* are often used to show percents. The whole pie is like the whole number one. If the whole pie is shaded, it represents 100%. If half of the pie is shaded, it represents one-half of 100% or 50%. Each shaded part of the pie illustrates a fraction or a percent of the whole. In the following examples, you will see data made into a visual picture with a pie graph.

Example 1

■ 25% of the book was illustrated with colored pictures.
■ 75% of the book was black print on a white background.

Example 2

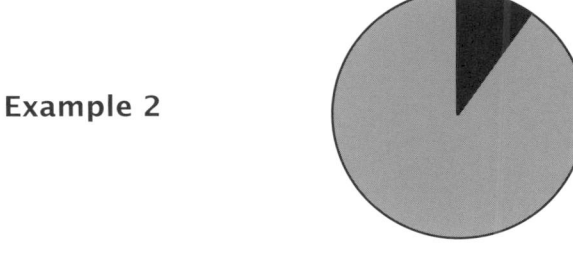

■ 10% of every paycheck goes to the church.
■ 90% of the paycheck is for our living expenses.

Example 3

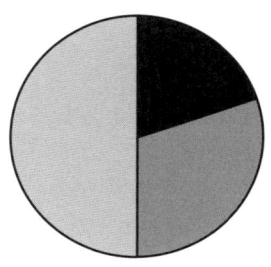

■ 20% of the noon meal was salad.
■ 30% of lunch was bread and rolls.
□ 50% of our lunch was soup.

Example 4

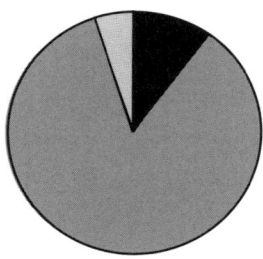

■ 10% of the population are left-handed.
■ 85% of the people are right-handed.
□ 5% are ambidextrous.

LESSON 14

Multiply All Decimals

What do you think will happen when you multiply by 1/1,000 or .001? The answer will move three places to the right, or the decimal point will appear to move three places to the left. This is because you are taking 1/1,000 of the original number. In the examples below, we'll estimate first by thinking what the answer should be, and then work the problem using both methods. The first method is to line up the decimal points as in an addition problem, and then figure out where to begin and start your progression from that point. In example 1, two-hundredths times nine-hundredths is 18 ten-thousandths. So we begin four spaces to the right.

The other method is to count the spaces to the right of the decimal point in both factors and add them for the answer. Then multiply the numbers with no thought of the decimal element. In the answer, count the spaces from the right to place the decimal point. Either method will work.

Example 1

```
    75.89              7589    two places
  ×   .52      or    ×   52  + two places
   1.578              1578
  37.945             37945
  39.4628           394628
                    39.4628   four places from the right
```

To estimate, round 75.89 to 80 and .52 to .5. Our estimation is 80 x .5 = 40. Or, think .5 is the same as five-tenths, which is one-half, and one-half of 80 is 40. The answer should be approximately 40.

Example 2

```
    536.042              536042    three places
  ×    .08      or   ×        08  + two places
    42.88336            4288336
                        42.88336  five places from the right
```

To estimate, round 536.042 to 500 and .08 to .1. Then 500 times .1 is 50. The answer should be approximately 50.

Example 3

```
      703.091                  703091    three places
  ×     1.584         or   ×      1584  + three places
      2.812364               2812364
     56.24728                5624728
    351.5455                3515455
    703.091                 703091
  1,113.696144             1113696144
                           1,113.696144  six places from the right
```

To estimate, round 703.091 to 700 and 1.584 to 2. Taking 700 times 2 gives 1,400. Or, 1.584 is close to 1.5 or 1 1/2, and 1 1/2 of 700 is 700 + 350 = 1,050. The answer should be approximately 1,050 or between 1,050 and 1,400, since 1.584 is between 1.5 and 2.

LESSON 15

Metric System Conversions – Part 2

Up to this point, the student has been asked to change from larger metric units to smaller metric units. When changing from larger to smaller, you are moving to the right in the values, and for each value, you increase by a factor of 10. Now the students are changing smaller to larger. I like to place the values with the larger one on the left, as in place value, before making the conversion. Changing from smaller to larger is like changing from pennies to dollars. Instead of increasing by a factor of 10 as we move each place to the right, we will do the opposite. For every place we move to the left (starting with the known value), we will decrease by a factor of 10, or divide by 10 for each place value.

Figure 1

			$1.00	$ .10	$ .01
kilo	hecto	deka	(unit)	deci	centi	milli
1,000	100	10	1	$\frac{1}{10}$	$\frac{1}{100}$	$\frac{1}{1,000}$

When moving along the decimal system from right to left, as when changing meters to kilometers, you divide because you know you are transforming a smaller unit to a larger unit of measure. Divide by 10 for each place value as you move to the left to make the number smaller.

But when changing hectometers to decimeters, or a larger to a smaller, you are breaking up a big piece into many smaller pieces, so you will multiply by powers of 10, or move the decimal point one place to the right (starting with the known value), for each place value. You know your number will be larger. Estimation is critical in this lesson. First, find out whether the number in the answer will be smaller or larger, and then count how many places.

Example 1
Convert: 5 dollars = ___ pennies.
(Estimate: should be more pennies than dollars.)

Since we have to transform dollars to pennies (large unit to small unit), we increase by a factor of 10 twice, or move the decimal point twice to make the result larger, so we get 500 pennies.

Example 2
Convert: ___ dollars = 700 pennies.
(Estimate: should be fewer dollars than pennies.)

Since we have to transform pennies to dollars (small unit to large unit), we decrease by a factor of 10 twice, or move the decimal point twice to make the result smaller, so we get 7 or $7.00.

Example 3
Convert: ___ hg = 560 cg
(Estimate: should be fewer hectograms than centigrams.)

We are changing centigrams to hectograms (small unit to large unit). So we divide by 10 four times, or move the decimal point four times to make the result smaller. The answer is .056 hg.

Example 4
Convert: 1.05 kg = ___ dg
(Estimate: should be more decigrams than kilograms.)

Since we have to transform kilograms to decigrams (large unit to small unit), we multiply by 10 four times, or move the decimal point to the right four times. The answer is 10,500 dg.

LESSON 16

Computing Area and Circumference
of a Circle with Pi = 3.14

The formula for the area of a circle is πr^2. *Pi*, or π, is a symbol for a value that is a little over three. The letter "r" represents the radius of a circle. The **radius** is the distance from the center of the circle to the edge of the circle. For help understanding and remembering the value of π, see figure 1. Whenever you're finding area, remember the word *squarea*, which is a conjoining of square and area. Area is always computed in square units. The value of π is normally represented by one of two values—the fractional value 22/7 or the decimal value 3.14. Both of these are approximations for a number that extends indefinitely as a decimal.
Pi = 3.1415927 . . .

There are some problems in which it is advantageous to use the decimal and others in which the fraction is more convenient. Since the focus of this book is decimals, we'll use 3.14 in the problems.

Figure 1

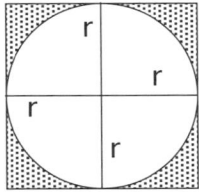 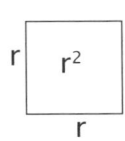 A little over three of the squares with a length of r and an area of r^2 equal the area of the circle.

Example 1

Find the area of the circle.

Area = πr^2

Area = $3.14(4.5)^2$

Area = $3.14(20.25)$

Area = 63.585 in^2

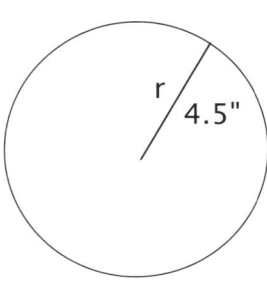

Example 2

Find the area of the circle.

Since the diameter is given, divide it by two to find the radius. (See example 5 for more on diameter.)

d = 2r, thus 13.6' = 2r, and 6.8' = r

Area = πr^2
Area = $3.14(6.8)^2$

Area = $3.14(46.24)$

Area = 145.1936 ft^2

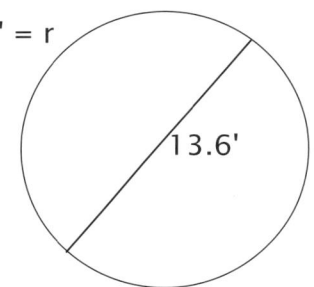

COMPUTING THE CIRCUMFERENCE

The formula for the circumference of a circle is $2\pi r$. Rectangles have area and perimeter, while circles have area and circumference. The *circumference* is the distance around the outside of a circle. A practical example of this is the length of a belt. It goes around your waist. Pi has the same value as before. The letter "r" still represents the radius of a circle. Circumference is always computed in linear units such as just plain inches or feet. Again, we'll use 3.14 for the value of π in all of our problems since this book is about decimals.

Example 3

Find the circumference of the circle.

Circumference = 2πr

Circumference = 2(3.14)(2.5)

Circumference = 15.7 in

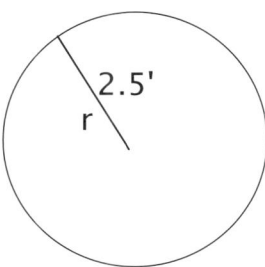

Example 4

Find the circumference of the circle.

Circumference = 2πr

Circumference = 2(3.14)(8.4)

Circumference = 52.752 ft

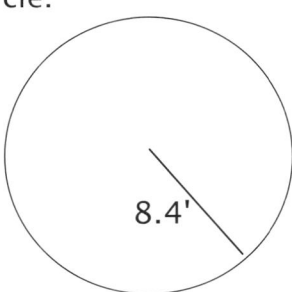

The distance from one edge of the circle to the other edge of the circle, through the center, is called the *diameter*. "Meter" means measure, and "dia" means through. The diameter is two times the radius. In example 4, the diameter is 2 x 8.4 ft or 16.8 ft.

In the formula for the circumference, 2πr, you can change the order of the three components to 2rπ. Since d (for diameter) is the same as twice the radius, then d = 2r. We can restate the formula as circumference = πd. Either formula is acceptable. Example 5 uses the new formula.

Example 5

Find the circumference of the circle.

Circumference = πd

Circumference = (3.14)(10)

Circumference = 31.4 in

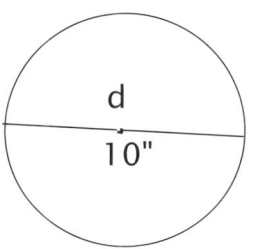

LESSON 17

Dividing a Decimal by a Whole Number

As in most math topics, there is how to do it and why you do it. When dividing a decimal by a whole number, put the decimal point directly above the decimal point inside the house, and then divide the numbers. The reason for this stems from multiplication. The missing factor on top of the house (known as the *quotient*), times the known factor (*divisor*) equals the product inside the house (*dividend*). Example 1 can be read as, "What number times 3 equals 1.2?" The answer, ".4," is a decimal. We could also read it as "Three times what number equals 1.2?" We know that .4 +. 4 + .4 =1.2, so 3 x (.4) = 1.2. If we place the decimal point in the quotient line directly above the decimal point in 1.2, and then divide 12 by 3, we get 4. Since the decimal point has already been placed there, it becomes .4. See example 1.

Notice in figures 1 and 2 how multiplication and division are related.

Figure 1

$$\text{factor} \overline{\smash{\big)}\,\text{product}}^{\text{factor}}$$

Figure 2

$$\text{divisor} \overline{\smash{\big)}\,\text{dividend}}^{\text{quotient}}$$

Example 1

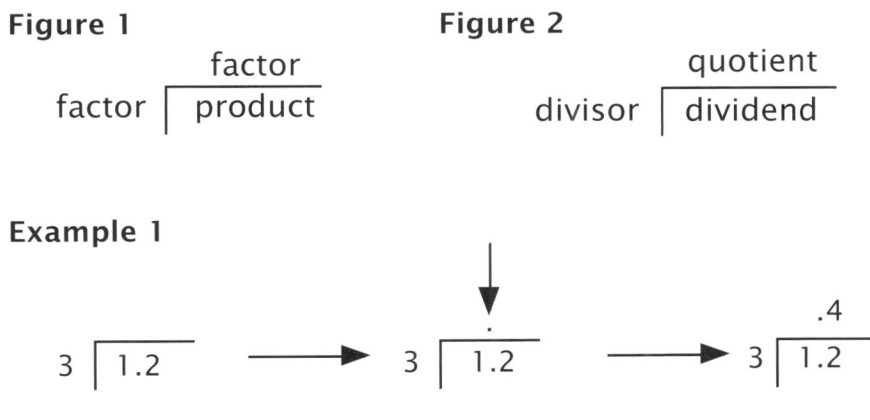

Always check your work: 3 x (.4) = 1.2

Example 2

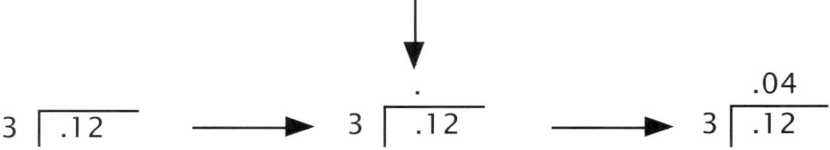

Always check your work: 3 x (.04) = .12

In the examples, notice that the division isn't difficult. It is where to place the values—or where to place the decimal point—that is tricky. Put the decimal point directly above the decimal point in the dividend, and then begin putting the missing factor directly above the number inside the house. The four is directly above the two in both examples, and the decimal points are directly above each other. However, in example 2 we needed a zero in the tenths place.

Example 3

Estimate: Seven is almost eight, so what times four is equal to eight? The answer is two.

```
    1.78
4 ⟌ 7.12
   −4
   ―――
    31
   −28
   ―――
     32
    −32
```

check
```
      4
   x 1.78
   ―――――
     .32
    2.8
    4.
   ―――――
    7.12
```

or

```
    1.78
   x  4
   ―――――
    7.12
```

The fraction 1/8 is the same as 1 ÷ 8. The number 1 is the same as 1.0 or 1.000. You can add zeros to a number as long as they are away from the decimal point and not in between the decimal point and a digit. The number 100.0 is not the same as 1.000. In example 4 we are going to change a fraction into a decimal by dividing one by eight.

Example 4

Estimate: Eight is almost 10, so what times 10 is equal to one? The answer is one tenth or .1.

```
      .125
   _____
 8 | 1.000
    -8
    ___
     20
    -16
    ___
     40
    -40
```

check

```
     8
  x .125
  _____
   .040
   .16
   .8
  _____
  1.000
```

or

```
   .125
   x 8
  _____
  1.000
```

Example 5

Estimate: 8.06 is almost 10, so what times five is equal to 10? The answer is two.

```
      1.612
   _____
 5 | 8.060
    -5
    ___
     30
    -30
    ___
      06
     -05
     ___
      10
     -10
```

check

```
       5
   x 1.612
   _____
    .010
    .05
    3.0
    5.
   _____
    8.060
```

or

```
   1.612
   x   5
   _____
   8.060
```

As in example 4, we needed to add another zero to 8.06. This is fine as long as the zero is outside or away from the decimal point. If the zero gets between another digit like 8 or 6 and the decimal point, it is wrong.

Example 6

Estimate: 200.07 is close to 200, and 9 is almost 10, so what times 10 is equal to 200?
The answer is 20.

```
        23.23
     _____
   9 ) 200.07
      −18
      ___
       20
      −18
      ___
        20
       −18
       ___
         27
        −27
```

check

```
         9
      x 22.23
       .27
       1.8
      18.
     180.
     _____
     200.07
```

or

```
      22.23
       x 9
      _____
      200.07
```

LESSON 18

Dividing a Whole Number by a Decimal
and Word Problems

There is a formula for dividing a whole number by a decimal. Let's think it through first, and then we'll see where the formula originates. We'll use money as our basis for this kind of problem. In example 1, the question is, "How many quarters, or how many groups of 25¢, are there in six dollars, or $6.00?" Remember that $.25 is a decimal, which is a fraction. So the answer should be larger, since you are asking how many fractions in a whole number.

Example 1

$$.25 \overline{\smash{)}6} = 24$$

If there are four quarters in one dollar, then there must be four times six or 24 quarters in six dollars.

The basis for the formula for dividing a whole number by a decimal is to make the divisor (the number outside the box) a whole number, so you are dividing a number by a whole number. To make .25 a whole number, we need to multiply it by 100, and 100 x .25 is 25. Recall that if we multiply the divisor by 100, we need to multiply the dividend by the same number as well, and 100 times 6 is 600. If we only multiplied the divisor by 10, it would be 2.5, which is still a decimal. See example 2 on the next page.

Example 2

$$25\overline{)600} \quad \begin{array}{r} 24 \\ \phantom{25\overline{)}}600 \\ -50 \\ \hline 100 \\ -100 \end{array}$$

.25 × 100 = 25 and 6 × 100 = 600

$$\frac{6}{.25} \times \frac{100}{100} = \frac{600}{25}$$

When we multiply both numbers by a multiple of 10, whether 10, 100, 1,000, etc., the decimal points appear to move the same number of spaces. Here is the formula: No matter how many spaces we have to move the decimal point to make the divisor (on the outside) a whole number, we move the decimal point on the inside (the dividend) the same number of places. See example 3.

Example 3

$$.25\overline{)6.00} \longrightarrow .25\overline{)6.00} \longrightarrow \begin{array}{r} 24. \\ 25\overline{)600} \\ -50 \\ \hline 100 \\ -100 \end{array} \quad \begin{array}{r} .25 \\ \times 24 \\ \hline 1.00 \\ 5.0 \\ \hline 6.00 \end{array}$$

In example 4 the question is, "How many sets of three dimes (.3), or how many groups of 30¢, are there in 18 dollars ($18.00)?" Remember that .3 and .30 are decimals, which are fractions. So, the answer should be larger than 18, since you are asking how many fractions are in a whole number.

Example 4

.3 ⟌ 18

Since there are a little over three sets of $.30 in one dollar, then there must be at least three times 18 or 54 sets in 18 dollars.

$$.3\overline{)18} \rightarrow .3\overline{)18.0} \rightarrow \begin{array}{r} 60. \\ .3\overline{)180} \\ -18 \\ \hline 00 \\ \underline{00} \end{array} \quad \begin{array}{r} 60 \\ \times\ .3 \\ \hline 18.0 \end{array}$$

60 is over 54, so it fits with our estimate, which is at least 54.

In example 5, the question is, "How many sets of five pennies (.05), or how many groups of 5¢, are there in 121 dollars ($121.00)?" Remember that .05 is a decimal, which is a fraction of a dollar.

Example 5

Since there are 20 nickels in a dollar, the answer should be 20 times larger than 121.

$$.05\overline{)121} \rightarrow .05\overline{)121.00} \rightarrow \begin{array}{r} 2{,}420. \\ 5.\overline{)12{,}100.} \\ -10\phantom{{,}000} \\ \hline 21 \\ -20 \\ \hline 10 \\ -10 \end{array}$$

20 times 121 is 2,420. So that works.

WORD PROBLEMS

Here are a few more multi-step word problems to try. You may want to read and discuss these with your student as you work out the solutions together. Again, the purpose is to stretch, not to frustrate. If you do not think the student is ready, you may want to come back to these later.

The answers are at the end of the solutions at the back of this book.

1. The boys ordered several pizzas for the weekend. When the first evening was over, the following amounts of pizza were left over: one-fourth of the pepperoni pizza, one-half of the cheese pizza, three-fourths of the mushroom pizza and one-fourth of the supreme pizza. The next morning, each boy ate the equivalent of one-fourth of a pizza for breakfast. If that finished the pizza, how many boys were there?

2. Dan read that an average snowfall of 10 inches yields one inch of water when melted. Very wet snow will measure five inches for one inch of water, and very dry snow may measure 20 inches for an inch of water. He made measurements for a storm that started with 5.3 inches of average snowfall. The precipitation changed to wet snow and dropped another 4.1 inches. The weather continued to warm up, and the storm finished with 1.5 inches of rain. What was the actual amount of water that fell during the storm? Round your answer to tenths.

3. Jim bought edging to go around a circular garden with a radius of three feet. Later he decided to double the diameter of the garden. How many more feet of edging must he buy?

4. One packet of flower seeds was enough to just fill the area of the smaller garden in #3. How many packets of seed are needed for the larger garden? Round each answer to the nearest square foot before continuing.

LESSON 19

Solving for an Unknown

When we have a problem like .3X = 12, and we are asked to solve for the unknown X, we need to get X by itself. We want our final step in a problem like this to be X = ___, with a number in the space. If X is multiplied by a number like .3, we need to make that number one. The reason is that one times X is still X.

Since X is multiplied by .3, we can use the inverse of multiplication, which is division, to make the coefficient one. (The *coefficient* is the number right before an unknown.) See example 1 for our solution.

Example 1

We divide both sides by .3, so the side on the left will be 1 · X or plain X.

$$.3X = 12$$
$$\frac{.3X}{.3} = \frac{12}{.3} \rightarrow .3\overline{)12} \rightarrow 3\overline{)120.}^{\,40.}$$
$$X = 40$$

Check the answer.   .3(40) = 12

It works.   12 = 12

Example 2

We divide both sides by .05, so the side on the left will be 1 · R or plain R.

$$.05R = 6$$

$$\frac{.05R}{.05} = \frac{6}{.05}$$

$$R = 120$$

$$.05\overline{)6} \rightarrow 5\overline{)600.}^{120.}$$

Check the answer. .05(120) = 6

It works. 6 = 6

LESSON 20

Dividing Decimal by a Decimal

Before I tell you how to do these problems, let's do one just by thinking. We'll use money for our example to make it clearer. See example 1. The question is, "How many nickels are there in 75 cents?"

Example 1

$$.05 \overline{\smash{)}.75}^{\,15}$$

From what we know of money, there are 15 nickels in 75¢. 15 × (.05) = .75

To do these problems, make the divisor, or the known factor, a whole number. To make .05 a whole number, we have to multiply it by 100. That makes a whole number, 5. We need to multiply *both* the divisor and the dividend by 100, or it changes the whole problem. Multiplying .75 times 100 gives 75. Our old problem in new clothes is 75 ÷ 5, as in example 2. We get the same answer both ways.

Example 2

$$\begin{array}{r} 15 \\ 5\overline{\smash{)}75} \\ \underline{-5} \\ 25 \\ \underline{-25} \end{array}$$

When we multiply both numbers by a multiple of 10, whether 10, 100, 1,000, etc., the decimal points appear to move the same number of spaces. So we could say that no matter how many spaces we have to move the decimal point to make the divisor (on the outside) a whole number, we move the decimal point on the inside (the dividend) the same number of places. See the following example. The number 3 is the same as 3.0.

Example 3

$$.03 \overline{)1.974} \longrightarrow .03 \overline{)1.974} \longrightarrow 3 \overline{)197.4}$$

```
                                              65.8
                                         3 ) 197.4
                                            −18
                                             17
                                            −15
                                             24
                                            −24
```

```
    .03              65.8
  × 65.8           ×  .03
  ─────           ──────
   .024            1.974
   .15
  1.80
  ─────
  1.974
```

Notice that in example 4 we need to add a zero to the dividend because there aren't enough places.

Example 4

$$.006 \overline{)63.87} \longrightarrow .006 \overline{)63.87} \longrightarrow 6 \overline{)63870}$$

```
                                           10645
                                        6 ) 63870
                                           −6
                                            3
                                           −0
                                           38
                                          −36
                                           −27
                                            24
                                            30
                                            30
```

```
   10645
 ×  .006
 ──────
  63.870
```

LESSON 21

Decimal Remainders

Remainders in a division problem involving decimals can be tricky. There are four possibilities, and I will go through each one with an accompanying example.

OPTION 1

With certain single-digit divisors like 2, 4, 5, or 8, if you add enough zeros onto the end of the dividend, eventually you will get the answer, or quotient, without a remainder.

Example 1
Divide 2.3 by 8.

```
        .2875
    8 | 2.3000
       -16
        70
       -64
         60
        -56
         40
        -40
```

OPTION 2

With other single-digit divisors like 3, 6, 7, or 9, you can add all the zeros you want and still never arrive at an answer without a remainder. It is with those problems in mind that we address options 2, 3, and 4. The most-used option is to select a place value that fits your needs and round to that value. A common value is the hundredths place. Before you begin dividing, simply state that you want the answer to be rounded to hundredths. This could be thousandths or millionths or any other place value; I am choosing hundredths. So all you need to do is divide to the thousandths place and then round to the hundredths place as in the example.

Example 2

Divide 29.6 by 7.

4.228 rounds to 4.23

```
      4.228
7 | 29.600
   -28
    ―――
    16
   -14
    ―――
     20
    -14
    ―――
      60
     -56
```

OPTION 3

Another option for divisors like 3, 6, 7, or 9 is to look for a pattern. We call these *repeating decimals*. When you divide by 3, eventually you will start seeing 3s in the quotient that go on and on. When you find a pattern like this, put a line above the repeating part to indicate that it goes on and on. Other divisors like 7 have very long numbers in their patterns before they repeat, as in example 3C. Study examples A, B, and C.

Example 3A

Divide 8.5 by 3.

$2.833\ldots$ equals $2.8\overline{3}$

```
     2.833
3 | 8.500
   -6
   ―――
    25
   -24
   ―――
     10
    - 9
    ―――
     10
    -9
```

Example 3B

Divide 4.6 by 9.

$5.111\ldots$ equals $5.\overline{1}$

```
      .5111
9 | 4.600
   -45
   ―――
    10
    -9
   ―――
     10
    - 9
    ―――
     10
    -9
```

Example 3C

Divide 9.5 by 7.

1.3$\overline{571428}$

571428 repeats.

Do you see it starting again, beginning with 57?

```
        1.357142857
      ┌──────────────
    7 │ 9.500000000
       -7
       ──
        25
       -21
       ───
         40
        -35
        ───
          50
         -49
         ───
           10
           -7
           ──
            30
           -28
           ───
             20
            -14
            ───
              60
             -56
             ───
               40
              -35
              ───
                50
               -46
```

OPTION 4

When you have a remainder in a decimal problem, and you don't want to round it off, but instead make your answer as accurate as possible, you can leave the answer with a fractional remainder. In example 4 on the next page, we divided to the hundredths place but wanted to stop there. So we put the 4 on top of the 7 because the next step in the problem is 4 divided by 7, and leave it like that. The answer is .27 4/7.

Remember though, that the 4/7 is really .04 divided by 7. When we subtract 14 from 1.93, the first number is really 1.4. The 53 is really .53 if we leave the decimal points in. When you check the problem, 7 times .27 is 1.89, and 7 times .04/7 is .04. Then .04 added to 1.89 is 1.93. So think of the 4/7 as occupying the same place as the 7 in .27.

Example 4

Divide 1.93 by 7.

$.27\frac{4}{7}$

$$\begin{array}{r} .27\frac{4}{7} \\ 7\overline{)1.93} \\ -14 \\ \overline{53} \\ -49 \\ \overline{4} \end{array}$$

Decimals with fractional remainders are most frequently used when the answer will be changed to a percent. For example, we might say that 33 1/3 percent of the people surveyed liked a certain product.

LESSON 22

More Solving for an Unknown

When we have a problem like .3X + .2 = 2, and we are asked to solve for the unknown X, we need to get X by itself. We want our final step in a problem like this to be X = ___, with a number in the space. We have learned how to make the coefficient one. Now we have to add to this knowledge how to make the number added to .3X (in the example + .2) a zero. We want the *variable*, or unknown, in example 1 to be alone on the left side of the equation and the numbers to be alone on the right side of the equation. The opposite, or inverse, of adding (+.2) is subtracting (+.2) from both sides. In example 1, we'll get all the numbers to one side, and then divide by .3 to make our variable just X.

Example 1

Subtract (.2) from both sides, so the side on the left has just the variable.

$$.3X + .2 = 2$$
$$\underline{-.2 = -.2}$$
$$.3X = 1.8$$
$$\frac{.3X}{.3} = \frac{1.8}{.3} \longrightarrow .3\overline{)1.8} \longrightarrow 3\overline{)18.}^{\,6.}$$

Divide both sides by .3, so the side on the left will be 1 · X or plain X.

$$X = 6$$

Check the answer. It works.

$$.3(6) + .2 = 2$$
$$1.8 + .2 = 2$$
$$2 = 2$$

Example 2

Subtract (3.3) from both sides, so the side on the left has just the variable.

$$1.5J + 3.3 = 13.8$$
$$-3.3 \quad -3.3$$
$$1.5J = 10.5$$

Divide both sides by 1.5, so the side on the left will be $1 \cdot J$ or plain J.

$$\frac{1.5J}{1.5} = \frac{10.5}{1.5} \longrightarrow 1.5\overline{)10.5} \longrightarrow 15\overline{)105.}^{\,7.}$$

$$J = 7$$

Check the answer.

$$1.5(7) + 3.3 = 13.8$$
$$10.5 + 3.3 = 13.8$$
$$13.8 = 13.8$$

It works.

Example 3

Subtract (.052) from both sides.

$$.12A + .052 = .1$$
$$-.052 \quad -.052$$
$$.12A = .048$$

Divide both sides by .12.

$$\frac{.12A}{.12} = \frac{.048}{.12} \longrightarrow .12\overline{).048} \longrightarrow 12\overline{)4.8}^{\,.4}$$

$$A = .4$$

Check the answer.

$$12(.4) + .052 = .1$$
$$.048 + .052 = .1$$
$$.1 = .1$$

It works.

LESSON 23

Transform Any Fraction
to a Decimal and a Percent

Recall that the line separating a fraction into a numerator and denominator also means "divided by." The fraction 1/4 means one divided by four. Now that we know how to divide numbers such as these, we can change any fraction into a decimal with hundredths. When dividing to make a percent, always stop at the hundredths place because that is what percents are—another way of writing hundredths. In examples 2–4, in order to make the answer stop at hundredths, we had to make the remainder a fraction. We shouldn't round it unless asked to, because that would not be accurate. Once we have changed to hundredths, it is one more step to percent. For more on remainders, see lesson 21.

Example 1
Transform 1/4 into a decimal and a percent.

$$\frac{1}{4} = 4 \overline{)1.00} \begin{array}{r} .25 \\ \underline{-8} \\ 20 \\ \underline{-20} \end{array} = 25\%$$

Example 2
Transform 2/3 into a decimal and a percent.

$$\frac{2}{3} = 3 \overline{)2.00} \begin{array}{r} .66 \\ \underline{-18} \\ 20 \\ \underline{-18} \\ 2 \end{array} = .\overline{6} \text{ or } .66\frac{2}{3} = 66\frac{2}{3}\%$$

Example 3

Transform 3/7 into a decimal and a percent.

$$\frac{3}{7} = 7 \overline{\smash{)}3.00}^{\;.42} = .42\frac{6}{7} = 42\frac{6}{7}$$
$$\phantom{\frac{3}{7} = 7)}\underline{-28}$$
$$\phantom{\frac{3}{7} = 7)\;}20$$
$$\phantom{\frac{3}{7} = 7)}\underline{-14}$$
$$\phantom{\frac{3}{7} = 7)\;\;}6$$

Example 4

Transform 5/8 into a decimal and a percent.

$$\frac{5}{8} = 8\overline{\smash{)}5.00}^{\;.62} = .62\frac{4}{8} = .62\frac{1}{2} = 62\frac{1}{2}$$

Example 5

Transform 1/3 into a decimal and a percent.

$$\frac{1}{3} = 3\overline{\smash{)}1.00}^{\;.33} = .33\frac{1}{3} = 33\frac{1}{3}\%$$

Earlier in the book, I suggested that the student memorize several fractions and their respective percents. The ones that were mentioned were 1/2, 1/4, 3/4, and 1/5. Of course, once you know one-fifth you can easily find the other fifths by multiplying by their numerator. So since one-fifth is 20%, two-fifths is 40%, etc.

There are other fractions that I suggest you also memorize. They are found in figure 1. You may find that in your world, other fractions are even more important than these, but the following fractions have helped me in everyday situations for years.

Figure 1

$$\frac{1}{3} = 33\frac{1}{3}\%$$

$$\frac{2}{3} = 66\frac{2}{3}\%$$

$$\frac{1}{7} = 14\frac{2}{7}\%$$

$$\frac{1}{8} = 12\frac{1}{2}\%$$

$$\frac{1}{9} = 11\frac{1}{9}\%$$

$$\frac{1}{20} = 5\%$$

A good application for this is finding your percentage score when grading a test. First write a fraction showing how many are correct. If you get 10 right out of 20, your score is 10/20 or 50%.

I usually use these when I can round them, since I am finding an approximate answer. So I think of one-third as 33%, two-thirds as 66%, one-ninth as 11%, etc. If I know one-ninth is approximately 11%, then five-ninths is five times that or 55%. The more you have tucked away upstairs (in your memory), the more you will have readily available to solve problems that arise unexpectedly. The more math you know, the more you can and will use.

LESSON 24

Decimals as Rational Numbers
and Word Problems

A *rational number* is a number that may be expressed as the result of dividing two numbers. A fraction, such as 7/8, is a rational number. Whole numbers may also be expressed as rational numbers, such as 5 = 5/1. Since decimals are fractions written in the base 10 system, they are a type of rational number. For example, to write the decimal .23 as a rational number, simply change it to 23/100. The numbers .23 and 23/100 are identical in value, but one is written as a rational number (or fraction), and the other is written in the base 10 system (or as a decimal). For help in learning what a rational number is, notice the first five letters in R-A-T-I-O-nal. Let me define a *ratio*.

Most of our study of fractions has been relegated to fractions of one. But in a ratio, the numerator and denominator are both whole numbers. For example, think of a classroom. If there are 10 boys and 15 girls, the ratio of boys to girls is 2 to 3, which can still be written as 2/3. Because of this ratio aspect, **ratio**nal numbers refer to fractions and are written as fractions, even though they represent whole numbers. Thus the word "rational" and how it relates to a ratio.

In the previous lesson, we transformed fractions into decimals (and then into percents). In this lesson, we will be doing the reverse and rewriting decimal numbers as reduced rational numbers or fractions.

Example 1
Transform .29 into a rational number.

$$.29 = \frac{29}{100}$$ You can't reduce this any further, so it is the final answer.

Example 2

Transform .8 into a rational number.

$$.8 = \frac{8 \div 2}{10 \div 2} = \frac{4}{5}$$

Example 3

Transform .036 into a rational number.

$$.036 = \frac{36 \div 4}{1000 \div 4} = \frac{9}{250}$$

Example 4

Transform .625 into a rational number.

$$.625 = \frac{625 \div 25}{1000 \div 25} = \frac{25 \div 5}{40 \div 5} = \frac{5}{8}$$

Example 5

Transform .035 into a rational number.

$$.035 = \frac{35 \div 5}{1000 \div 5} = \frac{7}{200}$$

WORD PROBLEMS

Here are a few more multi-step word problems to try. You may want to read and discuss these with your student as you work out the solutions together. Remember that the purpose is to stretch, not to frustrate. If you do not think the student is ready, you may want to come back to these later.

The answers are at the end of the solutions at the back of this book.

1. Emily cut two circles from a sheet of colored paper measuring 8 inches by 12 inches. One circle had a radius of 3 inches and the other had a radius of 2.5 inches. How many square inches of paper are left over? Is it possible to cut another circle with a 3-inch radius from the paper?

2. Tom wants to buy items costing $25.35, $50.69, and $85.96. He earns $6.50 an hour doing odd jobs. If 10 percent of his income is put aside for other purposes, how many hours must he work to earn the money he needs for his purchases? Round your answer to the nearest whole hour.

3. Three-tenths of the wooden toys were painted blue and one-fourth of them were painted green. Half of the remaining toys were painted red and half were painted yellow. If 300 toys are blue, how many are there of each of the other colors?

LESSON 25

Mean, Median, and Mode

In statistics, three categories are used to interpret data. I've been helped to remember which was which by key letters in each word. Use the blocks as we did when finding average, and then add these new words to clarify the data.

    MeAn = A represents the pure **A**verage, which we know how to do.

    MeDian = **MD** represents the **M**iddle number in the **D**ata, when arranged in ascending order.

    MOde = **MO** represents the number that occurs **M**ost **O**ften in the data.

Example 1

Interpret the following data, after arranging it in ascending order, to find the mean, median, and mode.

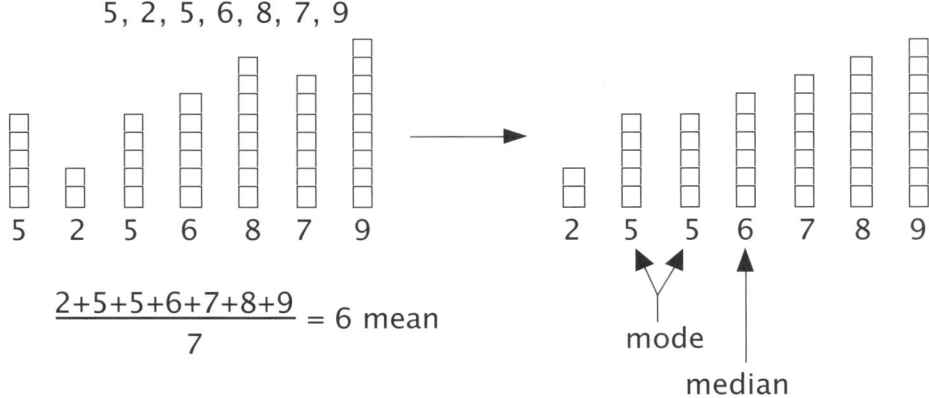

In example 1, the average, or mean, is six (all added up and divided by the number of items), the median is six (in the middle after arranging in ascending order), and the mode is five (appears twice or most often).

Example 2

Interpret the following data, after arranging it in ascending order, to find the mean, median, and mode: 10, 2, 9, 4, 10, 7

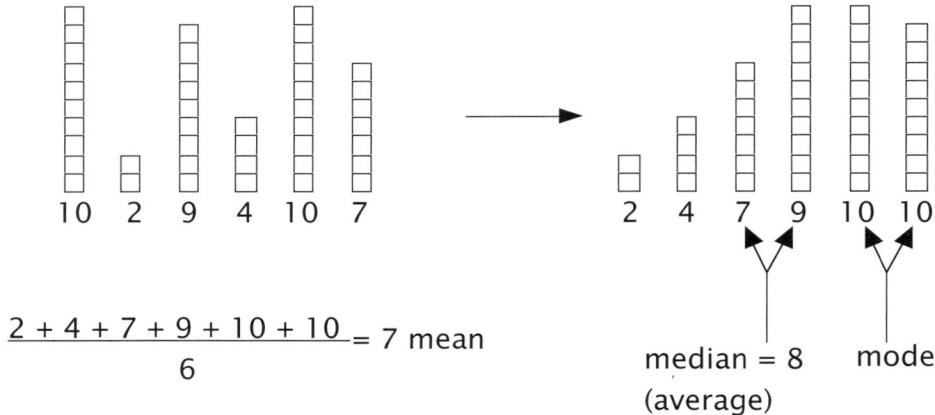

$$\frac{2 + 4 + 7 + 9 + 10 + 10}{6} = 7 \text{ mean}$$

median = 8 (average)   mode

The mean is 7 and the mode is 10. To find the median when there is an even number of data, find the average of the two middle ones. The median is 8.

LESSON 26

Probability

Probability is just like a fraction. The numerator is the **desired** outcome, and the denominator is all the **possible** outcomes. Think of probability as, "Out of all the possibilities, how many of the results are what I want?" Written as a fraction, it looks like this:

$$\frac{\text{Desired outcome}}{\text{Possible outcome}}$$

Example 1
You are rolling a six-sided die (singular of dice). On the sides are the numbers 1, 2, 3, 4, 5, and 6. Answer the following questions.

Question	Answer
What is the probability of rolling a 4?	$\dfrac{\text{Desired roll}}{\text{Possible rolls}} = \dfrac{4}{1-2-3-4-5-6} = \dfrac{1}{6}$
What is the probability of rolling a 5?	$\dfrac{\text{Desired roll}}{\text{Possible rolls}} = \dfrac{4}{1-2-3-4-5-6} = \dfrac{1}{6}$
What is the probability of rolling a 1, 3, or a 5 (an odd number)?	$\dfrac{\text{Desired roll}}{\text{Possible rolls}} = \dfrac{1-3-5}{1-2-3-4-5-6} = \dfrac{3}{6} = \dfrac{1}{2}$
What is the probability of rolling a 2, 4, or a 6 (an even number)?	$\dfrac{\text{Desired roll}}{\text{Possible rolls}} = \dfrac{2-4-6}{1-2-3-4-5-6} = \dfrac{3}{6} = \dfrac{1}{2}$
What is the probability of rolling a 1 or a 5?	$\dfrac{\text{Desired roll}}{\text{Possible rolls}} = \dfrac{1-5}{1-2-3-4-5-6} = \dfrac{2}{6} = \dfrac{1}{3}$

Example 2

You bought a book with 24 pages in it. Of the 24, 18 had printing, four had pictures, and two pages were blank. Answer the following questions.

Question	Answer
What is the probability of opening to a picture?	$\dfrac{\text{Desired pages}}{\text{Possible pages}} = \dfrac{4}{24} = \dfrac{1}{6}$
What is the probability of opening to printing?	$\dfrac{\text{Desired pages}}{\text{Possible pages}} = \dfrac{18}{24} = \dfrac{3}{4}$
What is the probability of opening to a blank page?	$\dfrac{\text{Desired pages}}{\text{Possible pages}} = \dfrac{2}{24} = \dfrac{1}{12}$
What is the probability of opening to pictures or printing?	$\dfrac{\text{Desired pages}}{\text{Possible pages}} = \dfrac{22}{24} = \dfrac{11}{12}$

LESSON 27

Points, Lines, Rays, and Line Segments

Geometry is the measure of the earth. "Geo" means earth and "metry" means measure. To measure the earth, we need to break it into smaller, more manageable pieces. The smallest unit of measure is an imaginary piece, called a ***point***. It has no measurable size, only position or location. We can't measure its width or length, so it has no dimensions or is zero-dimensional. To show something that is so small you really can't see it, we draw a dot. The dot is the "graph" of the point. It represents the point. We label it with a capital or uppercase letter. In figure 1, we call it point A.

Figure 1 • Point A

Using the point as the building block, consider a lot of connected points. A ***line*** is defined as an infinite (∞) number of connected points. It can be curved or straight. For our purposes, when we refer to a line it will be straight, unless mentioned otherwise. Since it is as wide as a point, it has no width, but it does have one dimension, which is length. So, a line is one-dimensional. A line is drawn or "graphed" with arrows at both ends to show that it goes on indefinitely or infinitely. It helps me to visualize a line as a laser beam. It is very, very thin but goes on and on. By our definition, a line is as thin as a point, but very long.

To label a line, use a lowercase letter, or choose two points (represented with uppercase letters) in the line. We call figure 2 "line m" and figure 3 "line QR" or $\overleftrightarrow{QR}$. In figure 3, the order of the points is not important. It could also be named "line RQ" or $\overleftrightarrow{RQ}$.

Figure 2

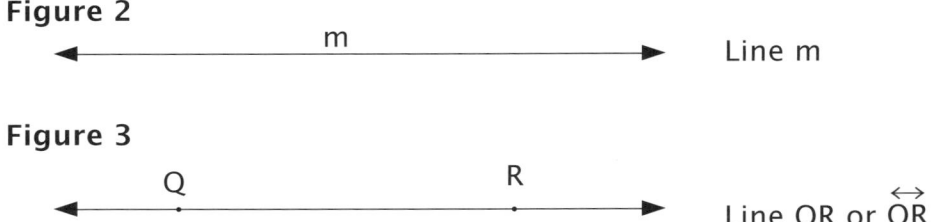

Line m

Figure 3

Line QR or $\overleftrightarrow{QR}$

A *ray* may be referred to as one-half of a line. It has a specific starting point at one end, called the endpoint or origin, and then proceeds infinitely in the other direction. Think of it as a flashlight or powerful ray that starts at one point and goes on and on in one direction. A line goes on in both directions, so it has arrows on both ends. A ray only has an arrow on one end. Figure 4 is labeled as $\overrightarrow{BC}$ and read as "ray BC." When you label a ray, the order of the points is very important. The first letter is always the origin or starting point. In our illustration, it is the flashlight or ray gun!

Figure 4

Ray $\overrightarrow{BC}$

A *line segment* is a finite, or measurable, piece of a line. It is not infinite, because it proceeds from a starting point to a stopping point. These two points are appropriately called endpoints. A line segment has a specific length. Figure 5 is labeled as $\overline{LH}$ and read as "line segment LH" or "segment LH." (The order of the letters is not important.)

Figure 5

Line Segment $\overline{LH}$

LESSON 28

Planes and Symbols

A *plane* is an infinite number of connected lines lying in the same flat surface. A plane has length and width, so it has two dimensions. It can be curved like a rolled-up piece of paper, but for our purposes it will be flat. Think of it as a floor or as a tabletop. It is long and wide and keeps going like a line in those two directions. But it is flatter than a pancake. It is thinner than a piece of paper because it is just as thin as a line, which is as thin as a point. We use a picture of a parallelogram to represent it, and we label it with a lowercase, or small, letter. In figure 1, the plane is referred to as "plane b."

Figure 1   b   Plane b

Most of our attention now will be on flat, two-dimensional shapes that lie in a plane. This is called ***plane geometry***. Three-dimensional geometry, with length, width, and height (or depth), pertaining to space and solids is called space or solid geometry. ***Solid geometry*** includes volume of a cube, cylinder, pyramid, cone, and sphere, all of which will be covered later.

To recap the lessons on geometry so far, consider that we operate in a three-dimensional place called space. The room you are in has length, width, and height. There are three dimensions. Look at figure 2 on the next page, and notice how we have progressed from no dimensions (a point) to one dimension (a line) to two dimensions (a plane), and finally to three dimensions.

Figure 2

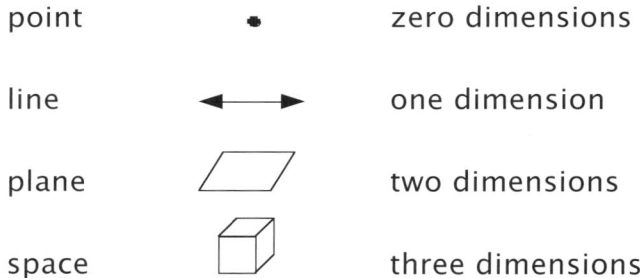

SYMBOLS

When speaking of two shapes being exactly the same shape and size, we say they are *congruent* (≅). The symbol comes from putting ~ and = together. The symbol "~" by itself means *similar,* or the same shape but not exactly the same size. Two squares have exactly the same shape but may have different measurements, and are said to be similar. Consider a house and a picture of a house. They have the same shape but are not the exact size, so they are similar.

The equal sign "=" means exactly the same length, or *equal*, and is used if two line segments are the same measurable length. Putting them together means exactly the same shape and size, which gives us congruent (≅). Use the equal sign for measurable objects with the same measure. Choose the congruent sign for objects that have the same shape and size.

Figure 3

Similar	~	For shapes
Equal	=	For numbers
Congruent	≅	For shapes

LESSON 29

Angles

If two lines intersect, the opening, or space between the lines, is referred to as an *angle*. In figure 1, there are four angles shown by the arcs. These are named 1, 2, 3, and 4.

Figure 1

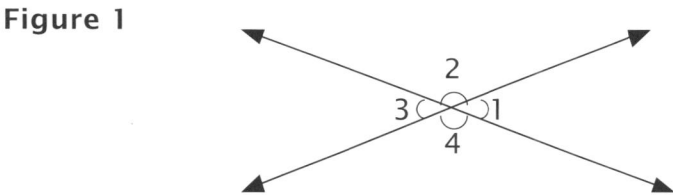

In figure 2, we are focusing on just one angle, made up by drawing two rays with the same endpoint. This endpoint, or origin, is called the *vertex*. (The plural of vertex is vertices.)

Figure 2

The rays are $\overrightarrow{DE}$ and $\overrightarrow{DF}$. The angle is labeled with either a number as in figure 1, or a lowercase Greek letter as in figure 2. We'll call this figure ∠α (angle alpha). Another way to identify this angle is by picking one point on each ray and the vertex, to make either ∠EDF or ∠FDE. Notice that the point labeling the vertex is always in the middle.

A *right angle* has a measure of 90 degrees (90°). It is the angle used most often in geometry. When the two rays or line segments make an angle with a measure of 90°, they form a right angle or square corner as in figure 3. Notice that it doesn't matter where the angle is, only that its measure is 90°. Usually, a box is used to represent the right angle.

There are 360 degrees in a circle. A right angle is thus one-fourth of a circle. See figure 4.

Figure 3

Figure 4

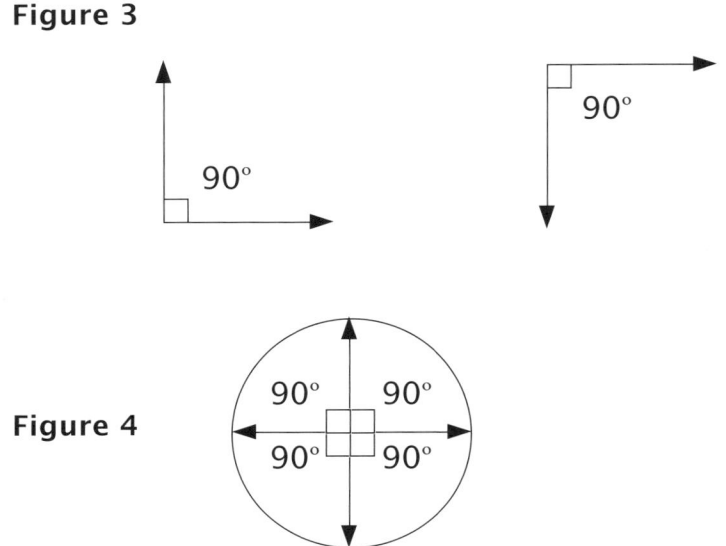

LESSON 30

Types of Angles

An *acute angle* is more than 0° but less then 90°. Most of the angles you see are acute angles.

Since they are small angles, it helps me to remember the name by thinking of "cute." Figure 1 has two acute angles.

Figure 1

0° < acute < 90°

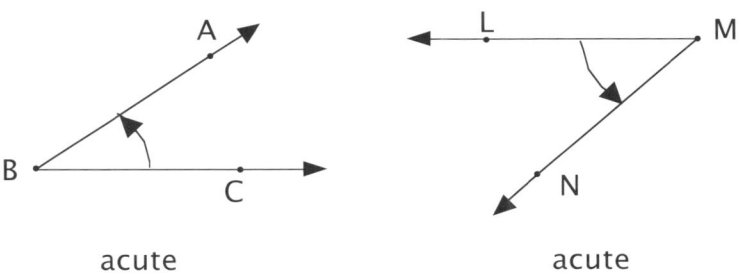

acute   acute

An *obtuse angle* is larger than 90° and less than 180°. See figure 2.

Figure 2

90° < obtuse < 180°

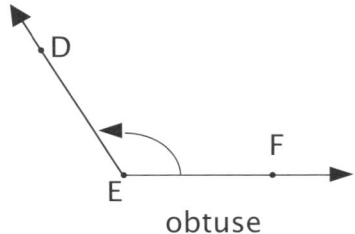   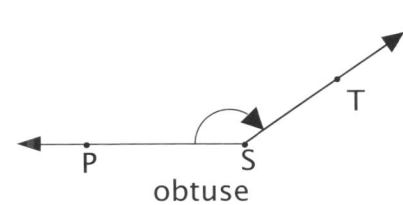

obtuse   obtuse

A *straight angle* is a lesser-known angle and is difficult to think of as an angle. It has a measure of 180°.

Figure 3

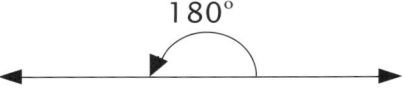

You sometimes hear of a car that skidded on ice and did a "one eighty," meaning it was going in one direction and then spun around until it was pointing in the opposite direction. This comes from the fact that it spun 180°. Figure 4 is a top view of this skid.

Figure 4

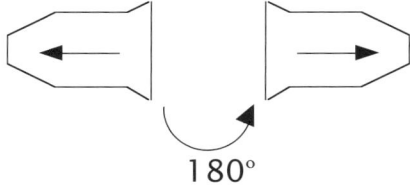

Student Solutions

Lesson Practice 1A
1. done
2. done
3. 4^2
4. 10^2
5. $5 \times 5 = 25$
6. $12 \times 12 = 144$
7. $4 \times 4 \times 4 = 64$
8. 6
9. $4 \times 4 = 16$
10. 100
11. done
12. 2^3
13. 5^2
14. 2^5
15. 8^2
16. 1^3
17. 4×4
18. 16

Lesson Practice 1B
1. 7^2
2. 12^2
3. 2^2
4. 6^2
5. $2 \times 2 \times 2 \times 2 = 16$
6. $1 \times 1 = 1$
7. $10 \times 10 = 100$
8. 8
9. $9 \times 9 = 81$
10. $3 \times 3 \times 3 = 27$
11. 2^1
12. 8^2
13. 2^4
14. 9^3
15. 6^4
16. 10^2
17. $5 \times 5 \times 5$
18. 125

Lesson Practice 1C
1. 5^2
2. 8^2
3. 10^2
4. 1^2
5. $1 \times 1 \times 1 = 1$
6. $11 \times 11 = 121$
7. $2 \times 2 = 4$
8. $3 \times 3 \times 3 \times 3 = 81$
9. $8 \times 8 = 64$
10. $6 \times 6 = 36$
11. 7^2
12. 4^3
13. 9^2
14. 2^3
15. 7^5
16. 3^3
17. 10×10
18. 100

Lesson Practice 1D
1. $7 \times 7 = 49$
2. $12 \times 12 = 144$
3. $100 \times 100 = 10,000$

LESSON PRACTICE 1D - SYSTEMATIC REVIEW 1F

4. $1 \times 1 \times 1 \times 1 = 1$
5. 9
6. $3 \times 3 = 9$
7. 10^2
8. 5^3
9. 3^3
10. 4^2
11. 8^4
12. 5^3
13. $1 \times 1 \times 1$
14. 1
15. done
16. $6 \div 3 = 2$
    $2 \times 1 = 2$
17. $20 \div 5 = 4$
    $4 \times 4 = 16$
18. $24 \div 6 = 4$
    $4 \times 5 = 20$
19. $\frac{1}{2}$ of 30 =
    $30 \div 2 = 15$
    $15 \times 1 = 15$ students
20. $\frac{7}{8}$ of 16 =
    $16 \div 8 = 2$
    $2 \times 7 = 14$ guests

11. 6^3
12. 9^2
13. 12×12
14. 144
15. $32 \div 8 = 4$
    $4 \times 3 = 12$
16. $12 \div 6 = 2$
    $2 \times 1 = 2$
17. $300 \div 3 = 100$
    $100 \times 2 = 200$
18. $72 \div 4 = 18$
    $18 \times 1 = 18$
19. $\frac{3}{4}$ of 600 =
    $600 \div 4 = 150$
    $150 \times 3 = 450$ visitors
20. $2 \times 12 = 24$ eggs
    $24 \div 6 = 4$
    $4 \times 1 = 4$ eggs

Systematic Review 1E
1. $2 \times 2 \times 2 = 8$
2. $4 \times 4 \times 4 \times 4 = 256$
3. $11 \times 11 = 121$
4. $8 \times 8 = 64$
5. $6 \times 6 \times 6 = 216$
6. 15
7. 12^2
8. 6^2
9. 10^3
10. 1^5

Systematic Review 1F
1. $9 \times 9 = 81$
2. $13 \times 13 = 169$
3. $3 \times 3 \times 3 \times 3 \times 3 = 243$
4. $5 \times 5 \times 5 \times 5 = 625$
5. 10
6. $20 \times 20 = 400$
7. 4^2
8. 7^3
9. 8^1
10. 3^4
11. 11^2
12. 10^3
13. $7 \times 7 \times 7 \times 7$
14. 2,401
15. $55 \div 5 = 11$
    $11 \times 2 = 22$
16. $210 \div 7 = 30$
    $30 \times 3 = 90$

SYSTEMATIC REVIEW 1F - LESSON PRACTICE 2C

17. $90 \div 10 = 9$
    $9 \times 1 = 9$
18. $54 \div 6 = 9$
    $9 \times 5 = 45$
19. $24 \div 6 = 4$
    $4 \times 1 = 4$ hours
20. $4 \times 60 = 240$ minutes

Lesson Practice 2A

1. 10^3 ; 10^2 ; 10^1 ; 10^0
2. $10 \times 10 \times 10 \times 10 \times 10 \times 10 = 1{,}000{,}000$
3. 10
4. $10 \times 10 \times 10 \times 10 = 10{,}000$
5. $10 \times 10 \times 10 \times 10 \times 10 = 100{,}000$
6. 10^4
7. 10^6
8. 10^1
9. 10^2
10. done
11. $2 \times 100 + 7 \times 10 + 6 \times 1$;
    $2 \times 10^2 + 7 \times 10^1 + 6 \times 10^0$
12. $1 \times 1{,}000 + 4 \times 100 + 9 \times 1$;
    $1 \times 10^3 + 4 \times 10^2 + 9 \times 10^0$
13. $3 \times 10{,}000 + 1 \times 1{,}000 + 5 \times 100$;
    $3 \times 10^4 + 1 \times 10^3 + 5 \times 10^2$
14. 8,403
15. 70,060
16. 4,962
17. 3,530
18. 52,174

Lesson Practice 2B

1. 10^3 ; 10^2 ; 10^1 ; 10^0
2. $10 \times 10 = 100$
3. $10 \times 10 \times 10 \times 10 \times 10 = 100{,}000$
4. $10 \times 10 \times 10 = 1{,}000$
5. $10 \times 10 \times 10 \times 10 \times 10 \times 10 = 1{,}000{,}000$
6. 10^0
7. 10^2
8. 10^4
9. 10^3
10. $4 \times 1{,}000 + 8 \times 100 + 3 \times 10 + 6 \times 1$;
    $4 \times 10^3 + 8 \times 10^2 + 3 \times 10^1 + 6 \times 10^0$
11. $6 \times 100{,}000 + 2 \times 100 + 7 \times 10 + 5 \times 1$
    $6 \times 10^5 + 2 \times 10^2 + 7 \times 10^1 + 5 \times 10^0$
12. $3 \times 100 + 8 \times 10 + 4 \times 1$;
    $3 \times 10^2 + 8 \times 10^1 + 4 \times 10^0$
13. 5×10;
    5×10^1
14. 9,349
15. 617
16. 40,703
17. 2,874
18. 12,211

Lesson Practice 2C

1. 10^3 ; 10^2 ; 10^1 ; 10^0
2. $10 \times 10 \times 10 \times 10 = 10{,}000$
3. $10 \times 10 = 100$
4. $10 \times 10 \times 10 \times 10 \times 10 \times 10 = 1{,}000{,}000$
5. 1
6. 10^5
7. 10^1
8. 10^3
9. 10^6
10. $7 \times 100 + 2 \times 1$;
    $7 \times 10^2 + 2 \times 10^0$
11. $1 \times 10{,}000 + 1 \times 1{,}000 + 6 \times 100 + 8 \times 1$;
    $1 \times 10^4 + 1 \times 10^3 + 6 \times 10^2 + 8 \times 10^0$
12. $8 \times 1{,}000{,}000$;
    8×10^6

LESSON PRACTICE 2C - SYSTEMATIC REVIEW 2F

13. $4 \times 10 + 8 \times 1$;
    $4 \times 10^1 + 8 \times 10^0$
14. 5,607
15. 1,980
16. 770,000
17. 3,610,000
18. 216,534

Systematic Review 2D
1. 10,000
2. 100
3. 1,000
4. 0
5. 2
6. 4
7. $3 \times 1,000 + 7 \times 100 + 6 \times 10 + 6 \times 1$;
   $3 \times 10^3 + 7 \times 10^2 + 6 \times 10^1 + 6 \times 10^0$
8. $5 \times 10,000 + 1 \times 1,000 + 1 \times 10 + 7 \times 1$;
   $5 \times 10^4 + 1 \times 10^3 + 1 \times 10^1 + 7 \times 10^0$
9. 6,220
10. 10,901
11. $1^2 = 1$
12. $2^3 = 8$
13. $2^4 = 16$
14. $\frac{1}{3} = \frac{2}{6} = \frac{3}{9} = \frac{4}{12}$
15. $\frac{2}{5} = \frac{4}{10} = \frac{6}{15} = \frac{8}{20}$
16. $42 \div 7 = 6$;
    $6 \times 5 = 30$ ducks
17. $12 \div 4 = 3$ months
    The letter is J
18. $24 \div 6 = 4$;
    $4 \times 5 = 20$ right-handed students
    $24 - 20 = 4$ left-handed students

Systematic Review 2E
1. 1
2. 1,000
3. 10
4. 1
5. 3
6. 0
7. $7 \times 100 + 4 \times 10 + 8 \times 1$;
   $7 \times 10^2 + 4 \times 10^1 + 8 \times 10^0$
8. $1 \times 10,000 + 2 \times 1,000 + 4 \times 100 + 6 \times 10 + 8 \times 1$;
   $1 \times 10^4 + 2 \times 10^3 + 4 \times 10^2 + 6 \times 10^1 + 8 \times 10^0$
9. 8,437
10. 60,294
11. $7^2 = 49$
12. $54^1 = 54$
13. $3^3 = 27$
14. $\frac{5}{6} = \frac{10}{12} = \frac{15}{18} = \frac{20}{24}$
15. $\frac{1}{4} = \frac{2}{8} = \frac{3}{12} = \frac{4}{16}$
16. $\frac{3}{8} = \frac{6}{16} = \frac{9}{24} = \frac{12}{32}$
17. $\frac{6}{7} = \frac{12}{14} = \frac{18}{21} = \frac{24}{28}$
18. $24 \div 12 = 2$;
    $2 \times 5 = 10$ hours
19. $12 \div 4 = 3$;
    $3 \times 1 = 3$ months
20. $28 \div 4 = 7$;
    $7 \times 3 = 21$ days

Systematic Review 2F
1. 100,000
2. 10,000
3. 100
4. 10^3
5. 10^4
6. 10^2

SYSTEMATIC REVIEW 2F - LESSON PRACTICE 3C

7. $5 \times 1,000 + 8 \times 100 + 8 \times 10 + 9 \times 1$;
   $5 \times 10^3 + 8 \times 10^2 + 8 \times 10^1 + 9 \times 10^0$
8. $6 \times 10,000 + 4 \times 100 + 1 \times 10$;
   $6 \times 10^4 + 4 \times 10^2 + 1 \times 10^1$
9. 7,260
10. 55,007
11. $9^2 = 81$
12. $1^5 = 1$
13. $2^4 = 16$
14. $\frac{2}{3} = \frac{4}{6} = \frac{6}{9} = \frac{8}{12}$
15. $\frac{3}{5} = \frac{6}{10} = \frac{9}{15} = \frac{12}{20}$
16. $\frac{1}{9} = \frac{2}{18} = \frac{3}{27} = \frac{4}{36}$
17. $\frac{7}{10} = \frac{14}{20} = \frac{21}{30} = \frac{28}{40}$
18. $60 \div 3 = 20$;
    $20 \times 2 = 40$ minutes
19. $75 \div 5 = 15$;
    $15 \times 4 = 60$ bean plants
20. $30 \div 10 = 3$ packages damaged
    $30 - 3 = 27$ packages arrived safely

Lesson Practice 3A

1. 10; 1; $\frac{1}{100}$; $\frac{1}{1,000}$
2. multiply
3. divide
4. done
5. $6 \times 10 + 7 \times 1 + 2 \times \frac{1}{10} + 1 \times \frac{1}{100}$
6. done
7. $2 \times 10^0 + 7 \times \frac{1}{10^2}$
8. 1,643.119
9. 371.045
10. 68.95
11. done
12. dollar; dimes; pennies; $1.00 + $.30 + $.08
13. dollar; dimes; pennies; $1.00 + $.70 + $.02
14. dime
15. penny

Lesson Practice 3B

1. 1,000; 100; 1; $\frac{1}{10}$; $\frac{1}{1,000}$
2. divide
3. multiply
4. $1 \times 1 + 9 \times \frac{1}{10} + 9 \times \frac{1}{100} + 9 \times \frac{1}{1,000}$
5. $2 \times 10 + 3 \times 1 + 6 \times \frac{1}{10} + 5 \times \frac{1}{100}$
6. $2 \times 10^2 + 3 \times 10^0 + 1 \times \frac{1}{10^2} + 6 \times \frac{1}{10^3}$
7. $8 \times 10^3 + 4 \times 10^2 + 3 \times 10^1 + 9 \times 10^0 + 7 \times \frac{1}{10^1}$
8. 2,401.613
9. 770.901
10. 5.008
11. dollars; dime; pennies; $.10; $.09; $2.19
12. 3; 2; 7; $3.00 + $.20 + $.07 = $3.27
13. dimes; pennies; $.40 + $.05 = $.45
14. penny
15. dime

Lesson Practice 3C

1. 1,000; 100; 10; $\frac{1}{10}$; $\frac{1}{100}$
2. left
3. right
4. $3 \times 100 + 4 \times 1 + 5 \times \frac{1}{100}$
5. $4 \times 1 + 6 \times \frac{1}{10} + 7 \times \frac{1}{100} + 9 \times \frac{1}{1,000}$
6. $6 \times 10^2 + 9 \times 10^1 + 1 \times 10^0 + 4 \times \frac{1}{10^1}$
7. $2 \times 10^1 + 5 \times 10^0 + 3 \times \frac{1}{10^1}$
8. 9,841.132
9. 3,006.084

LESSON PRACTICE 3C - SYSTEMATIC REVIEW 3F

10. 200.5
11. dollars; dimes; pennies
    $7.00 + $.40 + $.05 = $7.45
12. 1; 1; 4; $1.00 + $.10 + $.04 = $1.14
13. dimes; penny; $.60 + $.01 = $.61
14. dollar
15. dollar

Systematic Review 3D

1. $2 \times 10^3 + 3 \times 10^2 + 1 \times \dfrac{1}{10^1}$
2. $3 \times 10^1 + 8 \times 10^0 + 1 \times \dfrac{1}{10^1} + 2 \times \dfrac{1}{10^2} + 3 \times \dfrac{1}{10^3}$
3. 8,715.546
4. 6.411
5. dollars; dime; pennies; 3.00; .09; 3.19
6. 9; 4; 9.00 + .04 = 9.04
7. $5^2 = 25$
8. $1^5 = 1$
9. $10^3 = 1,000$
10. $10^0 = 1$
11. $\dfrac{1}{2} = \dfrac{2}{4} = \dfrac{3}{6} = \dfrac{4}{8}$
12. $\dfrac{5}{8} = \dfrac{10}{16} = \dfrac{15}{24} = \dfrac{20}{32}$
13. $\dfrac{8}{10} \div \dfrac{2}{2} = \dfrac{4}{5}$
14. $\dfrac{4}{24} \div \dfrac{4}{4} = \dfrac{1}{6}$
15. $\dfrac{6}{18} \div \dfrac{6}{6} = \dfrac{1}{3}$
16. $\dfrac{18}{30} \div \dfrac{6}{6} = \dfrac{3}{5}$
17. $4.00 + $.30 + $.06 = $4.36
18. $100 \div 10 = 10$;
    $10 \times 4 = 40$ cents

Systematic Review 3E

1. $6 \times \dfrac{1}{10^1} + 1 \times \dfrac{1}{10^2} + 5 \times \dfrac{1}{10^3}$
2. $1 \times 10^2 + 3 \times 10^1 + 5 \times 10^0 + 4 \times \dfrac{1}{10^3}$
3. 451.221
4. 10.607
5. dollar; dimes; pennies;
    $1.00 + $.00 + $.09 = $1.09
6. 6; 7; 5; $6.00 + $.70 + $.05 = $6.75
7. $3^2 = 9$
8. $4^3 = 64$
9. $2^2 = 4$
10. $10^3 = 1,000$
11. $\dfrac{4}{5} = \dfrac{8}{10} = \dfrac{12}{15} = \dfrac{16}{20}$
12. $\dfrac{3}{7} = \dfrac{6}{14} = \dfrac{9}{21} = \dfrac{12}{28}$
13. $\dfrac{11}{22} \div \dfrac{11}{11} = \dfrac{1}{2}$
14. $\dfrac{5}{25} \div \dfrac{5}{5} = \dfrac{1}{5}$
15. $\dfrac{4}{16} \div \dfrac{4}{4} = \dfrac{1}{4}$
16. $\dfrac{8}{32} \div \dfrac{8}{8} = \dfrac{1}{4}$
17. $1.00 + $.40 + $.07 = $1.47
18. $100 \div 5 = 20$;
    $20 \times 3 = 60$ cents
19. $\dfrac{2}{4} = \dfrac{1}{2}$ of a melon
20. $10^2 = 100$ blocks

Systematic Review 3F

1. $1 \times 10^0 + 1 \times \dfrac{1}{10^3}$
2. $1 \times 10^3 + 3 \times 10^2 + 5 \times 10^1 + 8 \times 10^0 + 9 \times \dfrac{1}{10^1} + 1 \times \dfrac{1}{10^2}$
3. 6,528.05
4. 2,000.986

5. dollars; dimes; pennies;
   $9.00 + $.80 + $.07 = $9.87
6. dollars; dimes; pennies;
   $2.00 + $.00 + $.08 = $2.08
7. $3^4 = 81$
8. $1^3 = 1$
9. $10^0 = 1$
10. $5^2 = 25$
11. $\frac{9}{10} = \frac{18}{20} = \frac{27}{30} = \frac{36}{40}$
12. $\frac{1}{6} = \frac{2}{12} = \frac{3}{18} = \frac{4}{24}$
13. $\frac{5}{30} \div \frac{5}{5} = \frac{1}{6}$
14. $\frac{14}{35} \div \frac{7}{7} = \frac{2}{5}$
15. $\frac{20}{40} \div \frac{20}{20} = \frac{1}{2}$
16. $\frac{18}{27} \div \frac{9}{9} = \frac{2}{3}$
17. $6.00 + $.09 = $6.09
18. $100 \div 100 = 1$;
    $1 \times 7 = 7$ cents
19. $\frac{9}{12} = \frac{3}{4}$ of the dishes
20. $20 \div 5 = 4$;
    $4 \times 4 = 16$ questions

Lesson Practice 4A

1. done
2. done
3.   1.53
     +1.12
      2.65
4.   2.17
     + .31
      2.48
5.   1.8
     +1.0
      2.8

6.   3.2
     + .4
      3.6
7.   1
      1.13
     +1.68
      2.81
8.   1
      1.67
     + .42
      2.09
9.   1.5
     +1.2
      2.7
10.   2.1
      + .8
       2.9
11.   1
       1.16
      +1.46
       2.62
12.   3.90
      + .02
       3.92
13.   1
       2.6
      +1.5
       4.1
14.   1
       1.8
      +1.3
       3.1
15.   3.00
      +1.62
       4.62
16.   4.48
      + .10
       4.58
17. $4.51
     +$ .35
      $4.86

LESSON PRACTICE 4A - LESSON PRACTICE 4C

18.
```
   1
   1.5
 +2.72
  4.22 miles
```

Lesson Practice 4B

1.
```
   1
   7.1
 + 6.2
  13.3
```

2.
```
   1
   5.9
 +1.2
   7.1
```

3.
```
    1
   2.45
 +5.07
   7.52
```

4.
```
    1
   4.13
 +1.96
   6.09
```

5.
```
   7.0
 +2.8
   9.8
```

6.
```
   1.5
 + 9.3
  10.8
```

7.
```
    1
   8.84
 + 3.09
  11.93
```

8.
```
   .437
 +.250
   .687
```

9.
```
   1
   8.8
 + 3.4
  12.2
```

10.
```
   6.2
 + .4
   6.6
```

11.
```
   1
   2.70
 + 9.41
  12.11
```

12.
```
   1
   5.52
 + .60
   6.12
```

13.
```
   3.9
 +4.0
   7.9
```

14.
```
   1
   7.5
 + .8
   8.3
```

15.
```
   4.15
 +3.00
   7.15
```

16.
```
   11
   .524
 +.277
   .801
```

17.
```
     1
   $12.95
   $ 15.50
   $28.45
```

18.
```
    1
   .625
 +2.125
   2.750  gallons
```

Lesson Practice 4C

1.
```
    1
   3.0
 + 9.8
  12.8
```

2.
```
   7.1
 +1.3
   8.4
```

LESSON PRACTICE 4C - SYSTEMATIC REVIEW 4D

3. 1 1
      1.95
    + 8.15
    ─────
    10.10

4.   1
      3.51
    +2.68
    ─────
    6.19

5.   1
      5.9
    + .4
    ─────
    6.3

6. 4.1
    +3.0
    ────
    7.1

7.   1
      2.34
    + .71
    ─────
    3.05

8. .440
    +.300
    ─────
    .740

9.   1
      6.5
    + 5.0
    ────
    11.5

10.   1
      2.8
    +5.9
    ────
    8.7

11. 1 1
    7.48
    +1.93
    ─────
    9.41

12. .162
    +8.000
    ──────
    8.162

13. 8.7
    + 8.1
    ────
    16.8

14. 6.0
    + .1
    ────
    6.1

15.   1
     .731
    + .402
    ─────
    1.133

16. 1.125
    + .112
    ─────
    1.237

17. 4.3
    + .5
    ────
    4.8 bushels

18. 2.045
    + .5
    ─────
    2.545 inches

Systematic Review 4D

1.   1
      1.5
    + 9.3
    ────
    10.8

2.   1
      5.9
    +1.6
    ────
    7.5

3. 6.34
    +2.41
    ─────
    8.75

4.   1
      1.82
    + 9.3
    ─────
    11.12

5. $2^3 = 8$

6. $6^2 = 36$

7. $10^4 = 10,000$

8. $7^2 = 49$

9. $1 \times 100 + 7 \times 10 + 6 \times 1 + 2 \times \frac{1}{10} + 1 \times \frac{1}{100}$

10. $6 \times \frac{1}{10} + 8 \times \frac{1}{100} + 5 \times \frac{1}{1,000}$

11. $4 \times 1 + 5 \times \dfrac{1}{10}$

12. $\dfrac{1}{4} = \dfrac{2}{8} = \dfrac{3}{12} = \dfrac{4}{16}$

13. $\dfrac{5}{8} = \dfrac{10}{16} = \dfrac{15}{24} = \dfrac{20}{32}$

14. $\dfrac{1}{4} + \dfrac{3}{5} = \dfrac{5}{20} + \dfrac{12}{20} = \dfrac{17}{20}$

15. $\dfrac{3}{4} + \dfrac{1}{6} = \dfrac{18}{24} + \dfrac{4}{24} = \dfrac{22}{24} = \dfrac{11}{12}$

16. $\dfrac{1}{3} + \dfrac{2}{5} = \dfrac{5}{15} + \dfrac{6}{15} = \dfrac{11}{15}$

17. $1\;1$
    75.25
    $+\;\;1.75$
    $\overline{77.00}$ inches

18. $12 \div 6 = 2$ spoiled apples
    $12 - 2 = 10$ good apples

Systematic Review 4E

1. $1\;1$
   8.6
   $+\;\;2.4$
   $\overline{11.0}$

2. 3.0
   $+4.4$
   $\overline{7.4}$

3. $1\;\;1$
   3.07
   $+\;\;9.25$
   $\overline{12.32}$

4. 5.00
   $+3.24$
   $\overline{8.24}$

5. $3^4 = 81$

6. $5^2 = 25$

7. $1^7 = 1$

8. $10^3 = 1{,}000$

9. $4 \times 10^1 + 3 \times 10^0 + 3 \times \dfrac{1}{10^1}$

10. $6 \times 10^0 + 1 \times \dfrac{1}{10^1} + 5 \times \dfrac{1}{10^3}$

11. $2 \times 10^2 + 3 \times \dfrac{1}{10^1} + 4 \times \dfrac{1}{10^2}$

12. $\dfrac{1}{2} = \dfrac{2}{4} = \dfrac{3}{6} = \dfrac{4}{8}$

13. $\dfrac{9}{10} = \dfrac{18}{20} = \dfrac{27}{30} = \dfrac{36}{40}$

14. $\dfrac{1}{9} + \dfrac{1}{2} = \dfrac{2}{18} + \dfrac{9}{18} = \dfrac{11}{18}$

15. $\dfrac{2}{5} + \dfrac{5}{6} = \dfrac{12}{30} + \dfrac{25}{30} = \dfrac{37}{30} = 1\dfrac{7}{30}$

16. $\dfrac{1}{10} + \dfrac{2}{3} = \dfrac{3}{30} + \dfrac{20}{30} = \dfrac{23}{30}$

17. $.5$
    $+.25$
    $\overline{.75}$ hours

18. $1\;1$
    9.5
    $+11.6$
    $\overline{21.1}$ gallons

19. $\dfrac{2}{3} + \dfrac{1}{5} = \dfrac{10}{15} + \dfrac{3}{15} = \dfrac{13}{15}$ of the problems

20. $30 \div 15 = 2$;
    $2 \times 13 = 26$ problems

Systematic Review 4F

1. 5.6
   $+4.3$
   $\overline{9.9}$

2. $1\;1$
   1.9
   $+\;\;9.2$
   $\overline{11.1}$

3. 1
   5.13
   $+\;\;9.50$
   $\overline{14.63}$

4. $1\;1$
   4.17
   $+1.95$
   $\overline{6.12}$

5. $8^2 = 64$

6. $10^0 = 1$

7. $4^3 = 64$

SYSTEMATIC REVIEW 4F - LESSON PRACTICE 5B

8. $9^2 = 81$
9. 9,500.1
10. 158.004
11. $\frac{1}{3} = \frac{2}{6} = \frac{3}{9} = \frac{4}{12}$
12. $\frac{3}{7} = \frac{6}{14} = \frac{9}{21} = \frac{12}{28}$
13. $\frac{2}{7} + \frac{1}{8} = \frac{16}{56} + \frac{7}{56} = \frac{23}{56}$
14. $\frac{3}{5} + \frac{2}{9} = \frac{27}{45} + \frac{10}{45} = \frac{37}{45}$
15. $\frac{3}{4} + \frac{1}{5} = \frac{15}{20} + \frac{4}{20} = \frac{19}{20}$
16.
```
      1
   $2.25
  +$1.69
   $3.94
```
17.
```
       1
   $ 4.00
   $ 2.50
  +$ 8.35
   $14.85
```
18. $\frac{5}{15} = \frac{1}{3}$
19. $\frac{3}{8} + \frac{1}{3} = \frac{9}{24} + \frac{8}{24} = \frac{17}{24}$; no
20. $27 \div 9 = 3$;
    $3 \times 5 = 15$ players

Lesson Practice 5A

1. done
2.
```
      7
    1⁄8.¹7
   − 4.8
    13.9
```
3.
```
    8  12
    9⁄.3⁄⁵
   −8.4 6
     .8 9
```

4.
```
    1 ¹¹
    2⁄2⁄.¹0
   − 9.6
    1 2.4
```
5.
```
    6.4
   −5.3
    1.1
```
6.
```
     4
    5⁄.¹0
   −2.4
    2.6
```
7.
```
    1   5
    2⁄3⁄.6⁄ ¹0
   − 9.4 3
    1 4.1 7
```
8.
```
        7
    2.8⁄ ¹1
   − .6 3
    2.1 8
```
9.
```
     3
    4⁄.¹0
   −2.6
    1.4
```
10.
```
     8
    9⁄.¹6
   − .9
    8.7
```
11.
```
    8.93
   −5.00
    3.93
```
12.
```
    3  16
    4⁄.7⁄ ¹0
   −1.9 8
    2.7 2
```
13.
```
    6  10
    $7⁄.1⁄ 7
   −$2.9 8
    $4.1 9
```

LESSON PRACTICE 5A - LESSON PRACTICE 5B

14.   5.75
     −1.50
     4.25 feet

15.   150.0
     − 4.5
     145.5 pounds

16.   15.15
     − 4.29
     10.86 seconds

17.   10.2
     − .7
     9.5 miles

18. $10.50
    −$ 4.79
     $ 5.71

Lesson Practice 5B

1.   8.3
    − .3
     8.0

2.   9.1
    −6.3
     2.8

3.   5.00
    −2.33
     2.67

4.   6.98
    −1.90
     5.08

5.   5.1
    −4.8
      .3

6.   9.2
    −1.0
     8.2

7.   8.60
    −5.35
     3.25

8.   12.75
    − 9.06
     3.69

9.   1.3
    − .6
      .7

10.   6.2
    −1.8
     4.4

11.   .513
    −.026
     .487

12.   .362
    −.100
     .262

13. $11.00
    −$ 5.75
     $ 5.25

14. $3.50
   −$2.45
     $1.05

15.   8.5
    −7.6
     .9 hours

LESSON PRACTICE 5B - SYSTEMATIC REVIEW 5E

16. 10
     11̸.¹0
    $-$ 4.5
     6.5 miles

17. 3 9
    12.4̸ 0̸¹0
    $-$ 2.375
     10.025 years

18. 1 10
     2̸.1̸¹0
    $-$ .16
     1.94 inches

Lesson Practice 5C

1. 7.8
   -1.2
    6.6

2. 3
    4̸.¹1
   -2.9
    1.2

3. 9 9
    1̸0̸.0̸¹0
   $-$ 2.67
    7.33

4. 4 9
    5̸.0̸¹1
   -1.92
    3.09

5. 5
    6̸.¹0
   $-$ .3
    5.7

6. 2
    3̸.¹7
   -2.8
    .9

7. 0 11
    11̸.2̸¹0
   $-$ 8.23
    2.97

8. 0 9
    1̸.0̸¹0
   $-$ .77
    .23

9. 8
    9̸.¹2
   $-$ .5
    8.7

10. 8.1
    -7.0
     1.1

11. .706
    $-.300$
     .406

12. 8
     .2 9̸ ¹8
    $-.009$
     .289

13. 1 4 9
    $\$ 2̸ 2 5̸.0̸¹0$
    $-\$$ 94.76
    $\$ 1 3 0.2 4$

14. 9 9
    $\$ 1̸0̸.0̸¹0$
    $-\$$ 7.45
    $\$$ 2.55

15. 6.7
    -4.3
     2.4 loaves

16. 1 9 9
    $\$ 2̸ 0̸.0̸¹0$
    $-\$ 1 6.25$
    $\$$ 3.75

17. $3\overset{210}{\cancel{1}.\cancel{0}}$
     -7.5
     23.5 chapters

18. $4\overset{3}{\cancel{5}}.6$
     -6.5
     39.1 ft

Systematic Review 5D

1. $\overset{3}{\cancel{4}}.\overset{1}{2}$
    -3.9
    $.3$

2. $8.\overset{5}{\cancel{6}}\overset{1}{0}$
    $-.04$
    8.56

3. $.007$
    $-.002$
    $.005$

4. $\overset{1}{.}99$
    $+.02$
    1.01

5. $\overset{1}{}5.7$
    $+2.3$
    8.0

6. 6.025
    $+.800$
    6.825

7. $\frac{1}{6} + \frac{4}{6} = \frac{5}{6}$

8. $\frac{2}{9} + \frac{1}{10} = \frac{20}{90} + \frac{9}{90} = \frac{29}{90}$

9. $\frac{2}{5} + \frac{3}{8} = \frac{16}{40} + \frac{15}{40} = \frac{31}{40}$

10. $1 \times 1{,}000 + 3 \times 100 + 4 \times \frac{1}{10}$

11. $1 \times 1 + 9 \times \frac{1}{10} + 7 \times \frac{1}{100} + 8 \times \frac{1}{1{,}000}$

12. done

13. $\frac{2}{3} - \frac{1}{4} = \frac{8}{12} - \frac{3}{12} = \frac{5}{12}$

14. $\frac{5}{6} - \frac{3}{7} = \frac{35}{42} - \frac{18}{42} = \frac{17}{42}$

15. $\overset{9}{\cancel{10}}.\overset{1}{2}5$
    -9.75
    $.50$ miles

16. $\$15.00 \$3\overset{5}{\cancel{6}}.\overset{16}{\cancel{7}}\overset{1}{5}$
    $+\$21.75 -\$31.99$
    $\$36.75 \$4.76$

17. $\frac{2}{3} - \frac{1}{6} = \frac{12}{18} - \frac{3}{18} = \frac{9}{18} = \frac{1}{2}$ of a pie

18. $\frac{1}{2} = \frac{2}{4}$ they got the same amount

Systematic Review 5E

1. $\overset{7}{\cancel{8}}.\overset{1}{1}6$
    -2.70
    5.46

2. $\overset{8}{\cancel{9}}.\overset{1}{0}4$
    -2.90
    6.14

3. $3.6\overset{3}{\cancel{4}}\overset{1}{3}$
    -2.008
    1.635

4. $\overset{1}{}6.9$
    $+1.2$
    8.1

5. 4.007
    $+.932$
    4.939

6. $\overset{1}{}31.600$
    $+.456$
    32.056

7. $\frac{1}{8} + \frac{7}{9} = \frac{9}{72} + \frac{56}{72} = \frac{65}{72}$

8. $\dfrac{4}{5} + \dfrac{2}{11} = \dfrac{44}{55} + \dfrac{10}{55} = \dfrac{54}{55}$

9. $\dfrac{1}{3} + \dfrac{1}{4} = \dfrac{4}{12} + \dfrac{3}{12} = \dfrac{7}{12}$

10. $2 \times 10^1 + 8 \times 10^0 + 7 \times \dfrac{1}{10^1} + 8 \times \dfrac{1}{10^2}$

11. 2×10^4

12. $\dfrac{8}{9} - \dfrac{5}{9} = \dfrac{3}{9} = \dfrac{1}{3}$

13. $\dfrac{4}{5} - \dfrac{1}{2} = \dfrac{8}{10} - \dfrac{5}{10} = \dfrac{3}{10}$

14. $\dfrac{7}{10} - \dfrac{1}{6} = \dfrac{42}{60} - \dfrac{10}{60} = \dfrac{32}{60} = \dfrac{8}{15}$

15.
```
    4
  7.5¹0
 -5.2 5
 ───────
  2.2 5 cups
```

16.
```
     1
    .90
   1.25
    .30
  + .50
  ──────
   2.95 miles
```

17.
```
     4.5
   +13.2
   ──────
    17.7 inches cut off
```

```
       13
    1  ¹
    2 4.¹0
   -1 7.7
   ────────
     6.3 inches left
```

18. $\dfrac{10}{100} = \dfrac{1}{10}$

19. $100 \div 10 = 10$;
    $10 \times 5 = 50$ cents

20. $\dfrac{2}{5} + \dfrac{1}{2} = \dfrac{4}{10} + \dfrac{5}{9} = \dfrac{9}{10}$ of them are gone
    $\dfrac{10}{10} - \dfrac{9}{10} = \dfrac{1}{10}$ still on the job

Systematic Review 5F

1.
```
    7
   8.¹30
  - 1.9
  ──────
    6.4
```

2.
```
    5
   6.¹18
  - 3.21
  ──────
    2.97
```

3.
```
     0 11
   7.1 2¹3
  -4.0 4 5
  ────────
   3.0 7 8
```

4.
```
     1
    6.1
  +  .9
  ──────
    7.0
```

5.
```
      1
    3.930
  +  .605
  ───────
    4.535
```

6.
```
    28.700
  +   .008
  ────────
    28.708
```

7. $\dfrac{2}{4} + \dfrac{2}{5} = \dfrac{10}{20} + \dfrac{8}{20} = \dfrac{18}{20} = \dfrac{9}{10}$

8. $\dfrac{2}{3} + \dfrac{1}{4} = \dfrac{8}{12} + \dfrac{3}{12} = \dfrac{11}{12}$

9. $\dfrac{1}{6} + \dfrac{1}{5} = \dfrac{5}{30} + \dfrac{6}{30} = \dfrac{11}{30}$

10. 150,000.04

11. 6,800.22

12. $\dfrac{5}{8} - \dfrac{1}{3} = \dfrac{15}{24} - \dfrac{8}{24} = \dfrac{7}{24}$

13. $\dfrac{9}{10} - \dfrac{2}{5} = \dfrac{45}{50} - \dfrac{20}{50} = \dfrac{25}{50} = \dfrac{1}{2}$

14. $\dfrac{1}{4} - \dfrac{1}{7} = \dfrac{7}{28} - \dfrac{4}{28} = \dfrac{3}{28}$

15.
```
     5 3
   6.¹3 4¹5
  -4.7 3 8
  ────────
   1.6 0 7 ounces
```

16.
```
   1 1
  $25.56
 +$ 6.78
  $32.34
```

17.
```
   2   2
  $ 3̷ 12.3̷ 14
  -$ 1 6.1 6
   $ 1 6.1 8
```

18. $100 \div 100 = 1$;
    $1 \times 17 = 17$ cents

19. $\frac{1}{3} + \frac{1}{4} = \frac{4}{12} + \frac{3}{12} = \frac{7}{12}$ of the candy

20. $36 \div 12 = 3$;
    $3 \times 7 = 21$ pieces given away
    $36 - 21 = 15$ pieces left

Lesson Practice 6A

1. b: meter
2. c: liter
3. a: gram
4. f: 10
5. d: 100
6. e: 1,000
7. $\frac{\text{kilogram(kg)}}{1,000\ g}$; $\frac{\text{hectogram(hg)}}{100\ g}$; $\frac{\text{dekagram(dkg)}}{10\ g}$; $\frac{\text{gram(g)}}{1\ g}$
8. $\frac{\text{kiloliter(kl)}}{1,000\ L}$; $\frac{\text{hectoliter(hl)}}{100\ L}$; $\frac{\text{dekaliter(dkl)}}{10\ L}$; $\frac{\text{liter(L)}}{1\ L}$
9. $\frac{\text{kilometer(km)}}{1,000\ m}$; $\frac{\text{hectometer(hm)}}{100\ m}$; $\frac{\text{dekameter(dkm)}}{10\ m}$; $\frac{\text{meter(m)}}{1\ m}$
10. 1,000 meters
11. 100 liters
12. 10 grams

Lesson Practice 6B

1. meter
2. liter
3. gram
4. hecto
5. deka
6. kilo
7. $\frac{\text{kilogram(kg)}}{1,000\ g}$; $\frac{\text{hectogram(hg)}}{100\ g}$; $\frac{\text{dekagram(dkg)}}{10\ g}$; $\frac{\text{gram(g)}}{1\ g}$
8. $\frac{\text{kiloliter(kl)}}{1,000\ L}$; $\frac{\text{hectoliter(hl)}}{100\ L}$; $\frac{\text{dekaliter(dkl)}}{10\ L}$; $\frac{\text{liter(L)}}{1\ L}$
9. $\frac{\text{kilometer(km)}}{1,000\ m}$; $\frac{\text{hectometer(hm)}}{100\ m}$; $\frac{\text{dekameter(dkm)}}{10\ m}$; $\frac{\text{meter(m)}}{1\ m}$
10. 100 grams
11. 1,000 liters
12. 10 meters

Lesson Practice 6C

1. c: meter
2. a: liter
3. b: gram
4. 1,000
5. 10
6. 100
7. $\frac{\text{kilogram(kg)}}{1,000\ g}$; $\frac{\text{hectogram(hg)}}{100\ g}$; $\frac{\text{dekagram(dkg)}}{10\ g}$; $\frac{\text{gram(g)}}{1\ g}$
8. $\frac{\text{kiloliter(kl)}}{1,000\ L}$; $\frac{\text{hectoliter(hl)}}{100\ L}$; $\frac{\text{dekaliter(dkl)}}{10\ L}$; $\frac{\text{liter(L)}}{1\ L}$
9. $\frac{\text{kilometer(km)}}{1,000\ m}$; $\frac{\text{hectometer(hm)}}{100\ m}$; $\frac{\text{dekameter(dkm)}}{10\ m}$; $\frac{\text{meter(m)}}{1\ m}$

10. 10 liters
11. 100 meters
12. 1,000 grams

14. $2\frac{3}{7} = \frac{17}{7}$

15. $5\frac{1}{3} = \frac{16}{3}$

16. $\begin{array}{r} \$5.95 \\ -\$1.50 \\ \hline \$4.45 \end{array}$

17. 1 km = 1,000 m

18. $\frac{7}{8} - \frac{1}{4} = \frac{28}{32} - \frac{8}{32} = \frac{20}{32} = \frac{5}{8}$ of a pizza

Systematic Review 6D

1. $\frac{\text{kilogram(kg)}}{1,000\text{ g}}$; $\frac{\text{hectogram(hg)}}{100\text{ g}}$; $\frac{\text{dekagram(dkg)}}{10\text{ g}}$; $\frac{\text{gram(g)}}{1\text{ g}}$

2. $\frac{\text{kiloliter(kl)}}{1,000\text{ L}}$; $\frac{\text{hectoliter(hl)}}{100\text{ L}}$; $\frac{\text{dekaliter(dkl)}}{10\text{ L}}$; $\frac{\text{liter(L)}}{1\text{ L}}$

3. $\frac{\text{kilometer(km)}}{1,000\text{ m}}$; $\frac{\text{hectometer(hm)}}{100\text{ m}}$; $\frac{\text{dekameter(dkm)}}{10\text{ m}}$; $\frac{\text{meter(m)}}{1\text{ m}}$

4. $\begin{array}{r} 1 \\ 1.8 \\ +4.5 \\ \hline 6.3 \end{array}$

5. $\begin{array}{r} 6.4 \\ -5.3 \\ \hline 1.1 \end{array}$

6. $\begin{array}{r} 1 \\ 7.23 \\ +3.54 \\ \hline 10.77 \end{array}$

7. $\begin{array}{r} 8.15 \\ -.13 \\ \hline 8.02 \end{array}$

8. $\frac{1}{6} + \frac{2}{3} = \frac{3}{18} + \frac{12}{18} = \frac{15}{18} = \frac{5}{6}$

9. $\frac{2}{5} + \frac{1}{2} = \frac{4}{10} + \frac{5}{10} = \frac{9}{10}$

10. $\frac{2}{3} - \frac{1}{4} = \frac{8}{12} - \frac{3}{12} = \frac{5}{12}$

11. $\frac{3}{5} - \frac{1}{3} = \frac{9}{15} - \frac{5}{15} = \frac{4}{15}$

12. $1\frac{5}{8} = \frac{13}{8}$

13. $4\frac{1}{2} = \frac{9}{2}$

Systematic Review 6E

1. $\frac{\text{kilogram(kg)}}{1,000\text{ g}}$; $\frac{\text{hectogram(hg)}}{100\text{ g}}$; $\frac{\text{dekagram(dkg)}}{10\text{ g}}$; $\frac{\text{gram(g)}}{1\text{ g}}$

2. $\frac{\text{kiloliter(kl)}}{1,000\text{ L}}$; $\frac{\text{hectoliter(hl)}}{100\text{ L}}$; $\frac{\text{dekaliter(dkl)}}{10\text{ L}}$; $\frac{\text{liter(L)}}{1\text{ L}}$

3. $\frac{\text{kilometer(km)}}{1,000\text{ m}}$; $\frac{\text{hectometer(hm)}}{100\text{ m}}$; $\frac{\text{dekameter(dkm)}}{10\text{ m}}$; $\frac{\text{meter(m)}}{1\text{ m}}$

4. $\begin{array}{r} 9.4 \\ +.5 \\ \hline 9.9 \end{array}$

5. $\begin{array}{r} 6 \\ \cancel{7}.^{1}0 \\ -3.1 \\ \hline 3.9 \end{array}$

6. $\begin{array}{r} 11 \\ 68.910 \\ +2.306 \\ \hline 71.216 \end{array}$

7. $\begin{array}{r} 4 \\ .4\cancel{5}\,^{1}3 \\ -.127 \\ \hline .326 \end{array}$

8. $\frac{2}{4} + \frac{1}{3} = \frac{6}{12} + \frac{4}{12} = \frac{10}{12} = \frac{5}{6}$

9. $\frac{2}{6} + \frac{1}{4} = \frac{8}{24} + \frac{6}{24} = \frac{14}{24} = \frac{7}{12}$

10. $\frac{3}{4} - \frac{1}{5} = \frac{15}{20} - \frac{4}{20} = \frac{11}{20}$

11. $\frac{4}{5} - \frac{1}{2} = \frac{8}{10} - \frac{5}{10} = \frac{3}{10}$

12. $2\frac{1}{3} = \frac{7}{3}$

13. $1\frac{1}{5} = \frac{6}{5}$

14. $6\frac{1}{2} = \frac{13}{2}$

15. $3\frac{4}{5} = \frac{19}{5}$

16. $5\frac{3}{4}$ dollars $= \frac{23}{4}$ dollars, or 23 quarters

17. $1 \times 10^2 + 7 \times 10^1 + 6 \times 10^0 + 4 \times \frac{1}{10^1}$

18. grams

19.
```
  11.50
 +12.25
 ──────
  23.75 minutes spent by Rachel
```

```
   2  9 14
   3̶ 0̶.5̶ 0̶
 - 2 3. 7 5
 ──────────
     6. 7 5 more minutes than Sierra
```

20.
```
  8.875
 -6.625
 ──────
  2.250 inches
```

Systematic Review 6F

1. $\frac{\text{kilogram(kg)}}{1{,}000 \text{ g}}$; $\frac{\text{hectogram(hg)}}{100 \text{ g}}$;
   $\frac{\text{dekagram(dkg)}}{10 \text{ g}}$; $\frac{\text{gram(g)}}{1 \text{ g}}$

2. $\frac{\text{kiloliter(kl)}}{1{,}000 \text{ L}}$; $\frac{\text{hectoliter(hl)}}{100 \text{ L}}$;
   $\frac{\text{dekaliter(dkl)}}{10 \text{ L}}$; $\frac{\text{liter(L)}}{1 \text{ L}}$

3. $\frac{\text{kilometer(km)}}{1{,}000 \text{ m}}$; $\frac{\text{hectometer(hm)}}{100 \text{ m}}$;
   $\frac{\text{dekameter(dkm)}}{10 \text{ m}}$; $\frac{\text{meter(m)}}{1 \text{ m}}$

4.
```
  6.11
 + .05
 ─────
  6.16
```

5.
```
  4.8
 -1.0
 ────
  3.8
```

6.
```
    1
  3.491
 +4.276
 ──────
  7.767
```

7.
```
    1 9 9
   2̶.0̶ 0̶ 17
 -  . 3 8 9
 ──────────
   1. 6 1 8
```

8. $\frac{2}{6} + \frac{1}{5} = \frac{10}{30} + \frac{6}{30} = \frac{16}{30} = \frac{8}{15}$

9. $\frac{1}{2} + \frac{3}{9} = \frac{9}{18} + \frac{6}{18} = \frac{15}{18} = \frac{5}{6}$

10. $\frac{4}{7} - \frac{1}{4} = \frac{16}{28} - \frac{7}{28} = \frac{9}{28}$

11. $\frac{5}{6} - \frac{1}{8} = \frac{40}{48} - \frac{6}{48} = \frac{34}{48} = \frac{17}{24}$

12. $1\frac{1}{8} = \frac{9}{8}$

13. $3\frac{2}{5} = \frac{17}{5}$

14. $5\frac{1}{4} = \frac{21}{4}$

15. $7\frac{3}{10} = \frac{73}{10}$

16. $3\frac{5}{6} = \frac{23}{6}$ of a pie; 23 people

17. $5^3 = 5 \times 5 \times 5 = 125$

18.
```
    1
  3.6
 + .8
 ────
  4.4 rolls used so far
```

```
   4
  5̶.10
 -4.4
 ────
   .6 rolls left
```

SYSTEMATIC REVIEW 6F - LESSON PRACTICE 7C

19. liters
20. $16 \div 4 = 4$;
    $4 \times 3 = 12$ children wanted to play
    $16 - 12 = 4$ children
    didn't want to play

Lesson Practice 7A

1. yard
2. quart
3. inch
4. f: $\frac{1}{10}$
5. e: $\frac{1}{100}$
6. d: $\frac{1}{1,000}$
7. $\frac{\text{gram}(g)}{1\,g}$, $\frac{\text{decigram}(dg)}{1/10\,g}$, $\frac{\text{centigram}(cg)}{1/100\,g}$, $\frac{\text{milligram}(mg)}{1/1,000\,g}$
8. $\frac{\text{liter}(L)}{1\,L}$, $\frac{\text{deciliter}(dl)}{1/10\,L}$, $\frac{\text{centiliter}(cl)}{1/100\,L}$, $\frac{\text{milliliter}(ml)}{1/1,000\,L}$
9. $\frac{\text{meter}(m)}{1\,m}$, $\frac{\text{decimeter}(dm)}{1/10\,m}$, $\frac{\text{centimeter}(cm)}{1/100\,m}$, $\frac{\text{millimeter}(mm)}{1/1,000\,m}$
10. 100
11. 1 deciliter
12. $28 \times 16 = 448$ grams

Lesson Practice 7B

1. gram
2. kilogram
3. kilometers
4. $\frac{1}{100}$
5. $\frac{1}{10}$
6. $\frac{1}{1,000}$
7. $\frac{\text{gram}(g)}{1\,g}$, $\frac{\text{decigram}(dg)}{1/10\,g}$, $\frac{\text{centigram}(cg)}{1/100\,g}$, $\frac{\text{milligram}(mg)}{1/1,000\,g}$
8. $\frac{\text{liter}(L)}{1\,L}$, $\frac{\text{deciliter}(dl)}{1/10\,L}$, $\frac{\text{centiliter}(cl)}{1/100\,L}$, $\frac{\text{milliliter}(ml)}{1/1,000\,L}$
9. $\frac{\text{meter}(m)}{1\,m}$, $\frac{\text{decimeter}(dm)}{1/10\,m}$, $\frac{\text{centimeter}(cm)}{1/100\,m}$, $\frac{\text{millimeter}(mm)}{1/1,000\,m}$
10. 1 milliliter
11. 1 centimeter
12. grams

Lesson Practice 7C

1. meters
2. 2 miles
3. milliliters
4. grams
5. 16 ounces
6. 36 inches
7. $\frac{\text{gram}(g)}{1\,g}$, $\frac{\text{decigram}(dg)}{1/10\,g}$, $\frac{\text{centigram}(cg)}{1/100\,g}$, $\frac{\text{milligram}(mg)}{1/1,000\,g}$
8. $\frac{\text{liter}(L)}{1\,L}$, $\frac{\text{deciliter}(dl)}{1/10\,L}$, $\frac{\text{centiliter}(cl)}{1/100\,L}$, $\frac{\text{milliliter}(ml)}{1/1,000\,L}$
9. $\frac{\text{meter}(m)}{1\,m}$, $\frac{\text{decimeter}(dm)}{1/10\,m}$, $\frac{\text{centimeter}(cm)}{1/100\,m}$, $\frac{\text{millimeter}(mm)}{1/1,000\,m}$
10. 1 millimeter
11. 1 liter
12. 50 grams

Systematic Review 7D

1. $\frac{\text{gram(g)}}{1\text{ g}}$, $\frac{\text{decigram(dg)}}{1/10\text{ g}}$, $\frac{\text{centigram(cg)}}{1/100\text{ g}}$, $\frac{\text{milligram(mg)}}{1/1{,}000\text{ g}}$
2. $\frac{\text{liter(L)}}{1\text{ L}}$, $\frac{\text{deciliter(dl)}}{1/10\text{ L}}$, $\frac{\text{centiliter(cl)}}{1/100\text{ L}}$, $\frac{\text{milliliter(ml)}}{1/1{,}000\text{ L}}$
3. $\frac{\text{meter(m)}}{1\text{ m}}$, $\frac{\text{decimeter(dm)}}{1/10\text{ m}}$, $\frac{\text{centimeter(cm)}}{1/100\text{ m}}$, $\frac{\text{millimeter(mm)}}{1/1{,}000\text{ m}}$
4. c: 1,000
5. b: 100
6. a: 10
7. $16 \div 2 = 8$
   $8 \times 1 = 8$
8. $40 \div 5 = 8$
   $8 \times 3 = 24$
9. $63 \div 9 = 7$
   $7 \times 2 = 14$
10. $48 \div 4 = 12$
    $12 \times 1 = 12$
11. $\frac{25}{8} = 3\frac{1}{8}$
12. $\frac{3}{2} = 1\frac{1}{2}$
13. $\frac{11}{9} = 1\frac{2}{9}$
14. $\frac{10}{3} = 3\frac{1}{3}$
15. liter
16. meter
17. 7.2
    $+6.3$
    $\overline{13.5}$ tons brought by the first two trucks

    17.80
    -13.50
    $\overline{4.30}$ tons needed
18. kilometers

Systematic Review 7E

1. $\frac{\text{gram(g)}}{1\text{ g}}$, $\frac{\text{decigram(dg)}}{1/10\text{ g}}$, $\frac{\text{centigram(cg)}}{1/100\text{ g}}$, $\frac{\text{milligram(mg)}}{1/1{,}000\text{ g}}$
2. $\frac{\text{liter(L)}}{1\text{ L}}$, $\frac{\text{deciliter(dl)}}{1/10\text{ L}}$, $\frac{\text{centiliter(cl)}}{1/100\text{ L}}$, $\frac{\text{milliliter(ml)}}{1/1{,}000\text{ L}}$
3. $\frac{\text{meter(m)}}{1\text{ m}}$, $\frac{\text{decimeter(dm)}}{1/10\text{ m}}$, $\frac{\text{centimeter(cm)}}{1/100\text{ m}}$, $\frac{\text{millimeter(mm)}}{1/1{,}000\text{ m}}$
4. deka
5. hecto
6. kilo
7. $30 \div 3 = 10$
   $10 \times 1 = 10$
8. $12 \div 6 = 2$
   $2 \times 5 = 10$
9. $49 \div 7 = 7$
   $7 \times 1 = 7$
10. $64 \div 8 = 8$
    $8 \times 5 = 40$
11. $\frac{37}{9} = 4\frac{1}{9}$
12. $\frac{17}{4} = 4\frac{1}{4}$
13. $\frac{47}{10} = 4\frac{7}{10}$
14. $\frac{53}{5} = 10\frac{3}{5}$
15. centimeters
16. 1,000
17. 1 mile
18. kilogram
19. 2.50
    $+3.25$
    $\overline{5.75}$ gallons used

    $6\overset{14}{\cancel{7}}.\cancel{5}0$
    -5.75
    $\overline{1.75}$ gallons left
20. $\frac{1}{2} - \frac{1}{6} = \frac{6}{12} - \frac{2}{12} = \frac{4}{12} = \frac{1}{3}$ of the job

Systematic Review 7F

1. $\frac{gram(g)}{1\ g}$; $\frac{decigram(dg)}{1/10\ g}$; $\frac{centigram(cg)}{1/100\ g}$; $\frac{milligram(mg)}{1/1{,}000\ g}$
2. $\frac{liter(L)}{1\ L}$; $\frac{deciliter(dl)}{1/10\ L}$; $\frac{centiliter(cl)}{1/100\ L}$; $\frac{milliliter(ml)}{1/1{,}000\ L}$
3. $\frac{meter(m)}{1\ m}$; $\frac{decimeter(dm)}{1/10\ m}$; $\frac{centimeter(cm)}{1/100\ m}$; $\frac{millimeter(mm)}{1/1{,}000\ m}$
4. dekagram
5. hectoliter
6. kilometer
7. $72 \div 8 = 9$
   $9 \times 3 = 27$
8. $50 \div 2 = 25$
   $25 \times 1 = 25$
9. $24 \div 3 = 8$
   $8 \times 2 = 16$
10. $44 \div 11 = 4$
    $4 \times 4 = 16$
11. $\frac{28}{6} = 4\frac{4}{6} = 4\frac{2}{3}$
12. $\frac{11}{7} = 1\frac{4}{7}$
13. $\frac{16}{5} = 3\frac{1}{5}$
14. $\frac{31}{2} = 15\frac{1}{2}$
15. the meter stick
16. 10 pounds
17. liters
18.
    ```
         1  1
      $  6.15
      $  8.45
     +$ 10.25
      $ 24.85
    ```
19. meters
20. $780 \div 10 = 78$
    $78 \times 3 = 234$ people

Lesson Practice 8A

1. done
2. 1,000 cm
3. 10 ml
4. 1,000 dl
5. 1,000 mg
6. 100 mg
7. done
8. 3,000,000 mm
9. 3,500 ml
10. 1,300 g
11. 600,000 cl
12. 1,000 dg
13. 60,000 cg
14. 100,000 dm
15. 90 mm

Lesson Practice 8B

1. 10,000 mm
2. 1,000,000 ml
3. 10 cl
4. 10 hg
5. 1,000 dm
6. 100 mm
7. 80,000 cm
8. 2,000 cl
9. 6,000,000 mg
10. 3,200 dg
11. 90 hl
12. 2,200,000 cm
13. 3,000 ml
14. 4,500 dg
15. 1,200 cm

Lesson Practice 8C

1. 100,000
2. 10
3. 10,000
4. 10,000

LESSON PRACTICE 8C - SYSTEMATIC REVIEW 8F

5. 10,000
6. 10
7. 1,000
8. 150,000
9. 700
10. 60
11. 5,000
12. 24,000
13. 500,000 seconds
14. 80 pieces
15. 2 liters = 2,000 ml
    2,000 > 1,983
    2 liters is more

Systematic Review 8D
1. 500 cg
2. 10,000 mm
3. 160,000 cm
4. c: 1,000
5. b: 100
6. a: 10
7. d: $\frac{1}{10}$
8. f: $\frac{1}{100}$
9. e: $\frac{1}{1,000}$
10. $2\frac{3}{8} = \frac{19}{8}$
11. $6\frac{1}{7} = \frac{43}{7}$
12. $\frac{17}{9} = 1\frac{8}{9}$
13. $\frac{8}{5} = 1\frac{3}{5}$
14. done
15. $3\frac{3}{8} + 1\frac{4}{5} = 3\frac{15}{40} + 1\frac{32}{40} = 4\frac{47}{40} =$
    $4 + \frac{40}{40} + \frac{7}{40} = 4 + 1 + \frac{7}{40} = 5\frac{7}{40}$
16. $2\frac{1}{10} + 3\frac{5}{8} = 2\frac{8}{80} + 3\frac{50}{80} = 5\frac{58}{80} = 5\frac{29}{40}$
17. 2,000 g
18. $1\frac{1}{2} + 1\frac{1}{4} = 1\frac{4}{8} + 1\frac{2}{8} = 2\frac{6}{8} = 2\frac{3}{4}$ hours

Systematic Review 8E
1. 32,000 ml
2. 90,000 ml
3. 150 mm
4. b: 1 yard
5. a: .4 of an inch
6. c: .6 of a mile
7. e: 1 quart
8. f: $\frac{1}{500}$ of a pound
9. d: 2.2 pounds
10. $5\frac{1}{8} = \frac{41}{8}$
11. $4\frac{2}{3} = \frac{14}{3}$
12. $\frac{22}{7} = 3\frac{1}{7}$
13. $\frac{27}{5} = 5\frac{2}{5}$
14. $8\frac{2}{3} + 5\frac{1}{4} = 8\frac{8}{12} + 5\frac{3}{12} = 13\frac{11}{12}$
15. $18\frac{1}{10} + 3\frac{5}{8} = 18\frac{8}{80} + 3\frac{50}{80} =$
    $21\frac{58}{80} = 21\frac{29}{40}$
16. $11\frac{3}{5} + 4\frac{1}{6} = 11\frac{18}{30} + 4\frac{5}{30} = 15\frac{23}{30}$
17. 100 cm = 10 dm; the same
18. 1 km
19. $\begin{array}{r} 3\;10\;14 \\ \cancel{4}\;\cancel{1}.\cancel{5}\;\cancel{0} \\ -\;3\;7.7\;5 \\ \hline 3.7\;5 \text{ inches} \end{array}$
20. $\frac{1}{3}$ of 900 = 300 in tanks
    $\frac{1}{6}$ of 900 = 150 in trucks
    300 + 150 = 450 riding
    900 − 450 = 450 walking
    You could also have added the fractions and used your answer to find the number of soldiers who were riding. There is often more than one way to solve a word problem.

Systematic Review 8F

1. 400,000 cg
2. 800 dkg
3. 5,000,000 ml
4. b: length
5. c: liquid
6. a: weight or mass
7. d: length
8. f: weight or mass
9. e: liquid
10. $2\frac{3}{5} = \frac{13}{5}$
11. $1\frac{1}{7} = \frac{8}{7}$
12. $\frac{19}{8} = 2\frac{3}{8}$
13. $\frac{31}{6} = 5\frac{1}{6}$
14. $9\frac{1}{3} + 6\frac{1}{4} = 9\frac{4}{12} + 6\frac{3}{12} = 15\frac{7}{12}$
15. $4\frac{2}{3} + 1\frac{1}{5} = 4\frac{10}{15} + 1\frac{3}{15} = 5\frac{13}{15}$
16. $12\frac{2}{10} + 4\frac{5}{8} = 12\frac{16}{80} + 4\frac{50}{80} =$
    $16\frac{66}{80} = 16\frac{33}{40}$
17. 45,000 g
    45,000,000 mg
18. $5 \times 10 = 50$ bottles
19. $1 \times 10^1 + 2 \times 10^0 + 8 \times \frac{1}{10^1} +$
    $7 \times \frac{1}{10^2} + 4 \times \frac{1}{10^3}$
20.
    ```
       6.20
       4.45
    +  9.00
    ─────────
      19.65 feet
    ```

Lesson Practice 9A

1. done
2. done

3.
```
   1+.4         1.4
 × 1+.2       ×1.2
 ────────    ──────
   .2+.08      .28
   1+.4        1.4
 ──────────  ──────
   1+.6+.08    1.68
```

4.
```
   1+.1        1.1
 ×   .6      ×  .6
 ────────    ──────
   .6+.06      .66
```

5.
```
   1+.2        1.2
 ×2+.1        ×2.1
 ────────    ──────
   .1+.02      .12
   2+.4        2.4
 ──────────  ──────
   2+.5+.02    2.52
```

6.
```
   2+.2        2.2
 ×   .3      ×  .3
 ────────    ──────
   .6+.06      .66
```

7.
```
   1+.3        1.3
 ×1+.1        ×1.1
 ────────    ──────
   .1+.03      .13
   1+.3        1.3
 ──────────  ──────
   1+.4+.03    1.43
```

8.
```
   2+.3        2.3
 ×   .2      ×  .2
 ────────    ──────
   .4+.06      .46
```

9.
```
   3.1
 ×  .3
 ──────
   .93 yards
```

10.
```
   2.1
 ×  .4
 ──────
   .84 gallons
```

Lesson Practice 9B

1.
```
   2+.8        2.8
 ×1+.0        ×1.0
 ────────    ──────
   .0+.00      .00
   2+.8        2.8
 ──────────  ──────
   2+.8+.00    2.80
```

2.
```
   2+.3        2.3
 ×   .1      ×  .1
 ────────    ──────
   .2+.03      .23
```

LESSON PRACTICE 9B - SYSTEMATIC REVIEW 9D

3.
```
  2 + .1        2.1
 ×1 + .4       ×1.4
  .8 + .04      .84
  2 + .1        2.1
  2 + .9 + .04  2.94
```

4.
```
  1 + .1        1.1
 ×    .8       ×  .8
  .8 + .08      .88
```

5.
```
  2 + .2        2.2
 ×1 + .2       ×1.2
  .4 + .04      .44
  2 + .2        2.2
  2 + .6 + .04  2.64
```

6.
```
  1 + .3        1.3
 ×    .3       ×  .3
  .3 + .09      .39
```

7.
```
  1 + .3        1.3
 ×1 + .3       ×1.3
  .3 + .09      .39
  1 + .3        1.3
  1 + .6 + .09  1.69
```

8.
```
  1 + .1        1.1
 ×    .9       ×  .9
  .9 + .09      .99
```

9.
```
  2.6
 ×1.1
  .26
  2.6
  2.86
```

10.
```
  1.1
 × .5
  .55 miles
```

3.
```
  1 + .3        1.3
 ×1 + .2       ×1.2
  .2 + .06      .26
  1 + .3        1.3
  1 + .5 + .06  1.56
```

4.
```
  2 + .1        2.1
 ×    .6       ×  .6
  1 + .2 + .06  1.26
```

5.
```
  2 + .1        2.1
 ×1 + .4       ×1.4
  .8 + .04      .84
  2 + .1        2.1
  2 + .9 + .04  2.94
```

6.
```
  2 + .1        2.1
 ×    .4       ×  .4
  .8 + .04      .84
```

7.
```
  2 + .3        2.3
 ×1 + .1       ×1.1
  .2 + .03      .23
  2 + .3        2.3
  2 + .5 + .03  2.53
```

8.
```
   .3          .3
  ×.3         ×.3
   .09         .09
```

9.
```
  3.2
 × .1
  .32 miles
```

10.
```
   .4
  ×.2
   .08 of the jelly beans
```

Lesson Practice 9C

1.
```
  1 + .3         1.3
 ×2 + .2        ×2.2
  .2 + .06       .26
  2 + .6         2.6
  2 + .8 + .06   2.86
```

2.
```
  2 + .0         2.0
 ×    .1        ×  .1
  .2 + .00       .2
```

Systematic Review 9D

1.
```
  2.5
 ×1.1
  .25
  2.5
  2.75
```

2.
```
  2.3
 ×1.3
  .69
  2.3
  2.99
```

3. 1.1
    $\underline{\times.5}$
    $.55$
4. $.2$
    $\underline{\times.2}$
    $.04$
5. 900 mg
6. 8,000 cl
7. 2,400 cg
8. cg
9. ml
10. km
11. dkl
12. dl
13. kg
14. done
15. done
16. $5\frac{1}{4} - 1\frac{2}{3} = 5\frac{3}{12} - 1\frac{8}{12} = 4\frac{15}{12} - 1\frac{8}{12} = 3\frac{7}{12}$
17. $3\frac{1}{2} - \frac{1}{3} = 3\frac{3}{6} - \frac{2}{6} = 3\frac{1}{6}$ yards
18. 3.3
    $\underline{\times.3}$
    $.99$ pies eaten

    ${}^2\cancel{3}^{12}$
    $\cancel{3}.\cancel{3}\,{}^{1}0$
    $\underline{-.99}$
    2.31 pies left over

Systematic Review 9E

1. 2.4
    $\underline{\times.2}$
    $.48$
2. 1.8
    $\underline{\times 1.1}$
    $.18$
    $\underline{1.8}$
    1.98
3. 1.7
    $\underline{\times.1}$
    $.17$

4. $.1$
    $\underline{\times.1}$
    $.01$
5. 280,000 cm
6. 3,600 dm
7. 500 dkl
8. meter
9. quart
10. gram
11. $7\frac{2}{5} - 3\frac{1}{3} = 7\frac{6}{15} - 3\frac{5}{15} = 4\frac{1}{15}$
12. $9 - 2\frac{1}{2} = 8\frac{2}{2} - 2\frac{1}{2} = 6\frac{1}{2}$
13. $6\frac{1}{2} - 4\frac{1}{7} = 6\frac{7}{14} - 4\frac{2}{14} = 2\frac{5}{14}$
14. $7\frac{1}{2} - 1\frac{1}{4} = 7\frac{2}{4} - 1\frac{1}{4} = 6\frac{1}{4}$ dollars or $6.25
    If one fraction can be made into an equivalent fraction with the same denominator as the other, you don't have to use the rule of four. The final result is the same.
15. 4.4
    $\underline{\times.1}$
    $.44$ pounds
16. ${}^1{}^1$
    45.15
    34.10
    $\underline{+7.05}$
    86.30 pounds
17. $6 \times 1,000 = 6,000$ steps
18. A meter is a little longer than a yard, so he ran faster in the 100 meter dash.

Systematic Review 9F

1. 2.3
    $\underline{\times.3}$
    $.69$
2. 2.1
    $\underline{\times 1.3}$
    $.63$
    $\underline{2.1}$
    2.73

SYSTEMATIC REVIEW 9F - LESSON PRACTICE 10B

4. .4
    × .2
    ―――
    .08

5. 140 mm
6. 2,000 g
7. 110 m
8. pounds
9. ounce
10. kilometer
11. $10\frac{1}{3} - 4\frac{1}{4} = 10\frac{4}{12} - 4\frac{3}{12} = 6\frac{1}{12}$
12. $6 - 1\frac{2}{3} = 5\frac{3}{3} - 1\frac{2}{3} = 4\frac{1}{3}$
13. $13\frac{1}{5} - 8\frac{5}{6} = 13\frac{6}{30} - 8\frac{25}{30} =$
    $12\frac{36}{30} - 8\frac{25}{30} = 4\frac{11}{30}$
14. 1 × 1,000 = 1,000 grams
15. 1,000 ÷ 1,000 = 1 kilogram
16. 1.1
    × .5
    ―――
    .55 pounds
17. 8 × 1,000 = 8,000 grams
18. $\overset{6}{\cancel{7}}.^{1}0$
    − 2 . 3
    ―――
    4 . 7 minutes

5. 4.7   47   1 place
    × .6   × 6 1 place
    ―――   ―――――
    2.82   2.82 2 places

6. .52   52   2 places
    ×.14   × 14 2 places
    ―――   ―――――
    .0208   208
    .052   52
    ―――   ―――――
    .0728   .0728 4 places

7. 200.   200   1 place
    × .08   × 08 1 place
    ―――   ―――――
    16.00   16.00 2 places

8. 5.28   528   2 places
    × .12   × 12 2 places
    ―――   ―――――
    .1056   1056
    .528   528
    ―――   ―――――
    .6336   .6336 4 places

9. $6.55 × .4 = $2.62
10. $.96 × 2.5 = $2.40
    (Note: although we don't usually include an ending 0 in decimals, it is customary to include hundredths when we are referring to money.)

Lesson Practice 10A

1. done
2. done
3. 35.   35   0 places
    × .26   × 26 2 places
    ―――   ―――
    2.10   2 10
    7.0   7 0
    ―――   ―――
    9.10   9.10 2 places

4. 1.03   103   2 places
    × .76   × 76 2 places
    ―――   ―――
    .0618   6 18
    .721   721
    ―――   ―――
    .7828   .7828 4 places

Lesson Practice 10B

1. 3.4   34   1 place
    × .94   × 94 2 places
    ―――   ―――
    .136   136
    3.06   306
    ―――   ―――
    3.196   3.196 3 places

2. .73   73   2 places
    ×.48   × 48 2 places
    ―――   ―――
    .0584   0584
    .292   292
    ―――   ―――
    .3504   .3504 4 places

3. 33.   33   0 places
    × .42   × 42 2 places
    ―――   ―――
    .66   66
    13.2   13 2
    ―――   ―――
    13.86   13.86 2 places

LESSON PRACTICE 10B - SYSTEMATIC REVIEW 10D

4.   5.19   5 19   2 places
      × .81   ×   8 1   2 places
      .0519   5 19
      4.152   4 152
      4.2039   4.2039   4 places

5.   9.6   96   1 place
      × .9   × 9   1 place
      8.64   8.64   2 places

6.   .64   64   2 places
      ×.94   × 94   2 places
      .0256   256
      .576   576
      .6016   .60 16   4 places

7.   116.   116   0 places
      × .02   × 02   2 places
      2.32   2.32   2 places

8.   7.18   718   2 places
      × .05   ×   5   2 places
      .3590   .3590   4 places

9. .15 × 100 = 15 ounces
10. .33 × 45 = 14.85 grams

Lesson Practice 10C

1.   4.8   48   1 place
      × .71   × 7 1   2 places
      .048   048
      3.36   3 36
      3.408   3.408   3 places

2.   .62   62   2 places
      ×.37   × 37   2 places
      .0434   434
      .186   186
      .2294   .2294   4 places

3.   69.   6 9   0 places
      × 2.3   × 23   1 place
      20.7   20 7
      138   138
      158.7   158.7   1 place

4.   9.97   997   2 places
      × .11   × 11   2 places
      .0997   997
      .997   997
      1.0967   1.0967   4 places

5.   1.7   17   1 place
      × .4   × 4   1 place
      .68   .68   2 places

6.   .16   16   2 places
      ×.54   × 54   2 places
      .0064   64
      .080   80
      .0864   .0864   4 places

7.   400.   4 00   0 places
      × .11   × 11   2 places
      44.00   44.00   2 places

8.   6.73   673   2 places
      × .46   × 46   2 places
      .4038   4038
      2.692   2 692
      3.0958   3.0958   4 places

9. 4.25 × .75 = 3.1875 bushels
10. 2.45 × 5 = 12.25 meters

Systematic Review 10D

1.   2.6   26   1 place
      × .24   × 24   2 places
      .104   104
      .52   52
      .624   .624   3 places

2.   6.3   63   1 place
      × 5.7   × 57   1 place
      4.41   4 41
      31.5   3 15
      35.91   35.91   2 places

3.   3.52   352   2 places
      × .04   × 04   2 places
      .1408   .1408   4 places

SYSTEMATIC REVIEW 10D - LESSON PRACTICE 11A

4. $.67$ 67 2 places
    $\underline{\times .05}$ $\underline{\times 05}$ 2 places
    $.0335$ $.0335$ 4 places

5. 14,000 dg
6. 50 cm
7. $1.00 - .12 = .88$
8. $4.08 - 2.9 = 1.18$
9. $.95 + 3.61 = 4.56$
10. $4\frac{3}{4} + 2\frac{1}{5} = 4\frac{15}{20} + 2\frac{4}{20} = 6\frac{19}{20}$
11. $12\frac{4}{7} - 6\frac{1}{7} = 6\frac{3}{7}$
12. $5\frac{1}{10} + 3\frac{2}{5} = 5\frac{5}{50} + 3\frac{20}{50} =$
    $8\frac{25}{50} = 8\frac{1}{2}$
    or
    $5\frac{1}{10} + 3\frac{4}{10} = 8\frac{5}{10} = 8\frac{1}{2}$
13. $\frac{1}{\cancel{2}} \times \frac{\cancel{2}}{3} = \frac{1}{3}$
14. $\frac{3}{4} \times \frac{1}{\cancel{9}_3} = \frac{1}{12}$
15. $\frac{\cancel{6}^3}{8} \times \frac{1}{\cancel{2}} = \frac{3}{8}$
16. $\frac{2}{3} \times \frac{1}{\cancel{2}} = \frac{1}{3}$ of the package

    Writing "1" when canceling is optional.
17. $.4 \times 18 = 7.2$ in
18. $55 \times 2.2 = 121$ lb

Systematic Review 10E

1. 11.9 119 1 place
    $\underline{\times 5.3}$ $\underline{\times 53}$ 1 place
    3.57 357
    $\underline{59.5}$ $\underline{595}$
    63.07 63.07 2 places

2. $.05$ 05 2 places
    $\underline{\times .4}$ $\underline{\times 4}$ 1 place
    $.02$ $.020$ 3 places

3. 1.07 107 2 places
    $\underline{\times .72}$ $\underline{\times 72}$ 2 places
    $.0214$ 214
    $\underline{.749}$ $\underline{749}$
    $.7704$ $.7704$ 4 places

4. $.46$ 46 2 places
    $\underline{\times .49}$ $\underline{\times 49}$ 2 places
    $.0414$ 414
    $\underline{.184}$ $\underline{184}$
    $.2254$ $.2254$ 4 places

5. 1,700 dg
6. 800,000 mm
7. $12.51 - 2.74 = 9.77$
8. $2.38 + 4.509 = 6.889$
9. $.412 - .367 = .045$
10. $5\frac{3}{7} + 5\frac{1}{4} = 5\frac{12}{28} + 5\frac{7}{28} = 10\frac{19}{28}$
11. $2\frac{1}{8} - 1\frac{5}{9} = 2\frac{9}{72} - 1\frac{40}{72} = 1\frac{81}{72} - 1\frac{40}{72} = \frac{41}{72}$
12. $3\frac{1}{2} + 2\frac{1}{2} = 5\frac{2}{2} = 6$
13. $\frac{1}{2} \times \frac{3}{5} = \frac{3}{10}$
14. $\frac{3}{4} \times \frac{1}{7} = \frac{3}{28}$
15. $\frac{\cancel{4}^2}{5} \times \frac{7}{\cancel{10}_5} = \frac{14}{25}$
16. $\frac{3}{4} \div \frac{6}{1} = \frac{3}{4} \times \frac{1}{\cancel{6}_2} = \frac{1}{8}$
17. $1\frac{1}{6} + 5\frac{3}{8} = 1\frac{8}{48} + 5\frac{18}{48} =$
    $6\frac{26}{48} = 6\frac{13}{24}$ yd
18. $2.5 \times 36 = 90$ cm
19. $1.06 \times 8 = 8.48$ qt
20. 34,000 grams

Systematic Review 10F

1. 9.35 935 2 places
    $\underline{\times .5}$ $\underline{\times 5}$ 1 place
    4.675 4.675 3 places

Systematic Review 10F

2.
$$\begin{array}{r} 8.9 \\ \times\ 8.9 \\ \hline 8.01 \\ 71.2 \\ \hline 79.21 \end{array}$$
$$\begin{array}{r} 89 \quad \text{1 place} \\ \times\ 89 \quad \text{1 place} \\ \hline 801 \\ 712 \\ \hline 79.21 \quad \text{2 places} \end{array}$$

3.
$$\begin{array}{r} 6.31 \\ \times\ .73 \\ \hline .1893 \\ 4.417 \\ \hline 4.6063 \end{array}$$
$$\begin{array}{r} 631 \quad \text{2 places} \\ \times\ 73 \quad \text{2 places} \\ \hline 1893 \\ 4\ 4\ 17 \\ \hline 4.6063 \quad \text{4 places} \end{array}$$

4.
$$\begin{array}{r} .29 \\ \times .34 \\ \hline .0116 \\ .087 \\ \hline .0986 \end{array}$$
$$\begin{array}{r} 29 \quad \text{2 places} \\ \times\ 34 \quad \text{2 places} \\ \hline 116 \\ 87 \\ \hline .0986 \quad \text{4 places} \end{array}$$

5. 40 dkl
6. 30 L
7. $1.26 - .63 = .63$
8. $3.052 + 1.98 = 5.032$
9. $400 - .5 = 399.5$
10. $6\frac{3}{6} + 8\frac{2}{5} = 6\frac{15}{30} + 8\frac{12}{30} = 14\frac{27}{30} = 14\frac{9}{10}$
11. $3 - 2\frac{1}{4} = 2\frac{4}{4} - 2\frac{1}{4} = \frac{3}{4}$
12. $7\frac{3}{10} + 9\frac{4}{5} = 7\frac{15}{50} + 9\frac{40}{50} = 16\frac{55}{50} = 17\frac{5}{50} = 17\frac{1}{10}$
13. $\frac{3}{\cancel{2}\ 8} \times \frac{\cancel{4}}{7} = \frac{3}{14}$
14. $\frac{1}{2} \times \frac{\cancel{8}^{4}}{9} = \frac{4}{9}$
15. $\frac{3}{5} \times \frac{3}{5} = \frac{9}{25}$
16. $\frac{\cancel{2}}{3} \times \frac{1}{\cancel{2}} = \frac{1}{3}$
17. $4 - 1\frac{2}{3} = 3\frac{3}{3} - 1\frac{2}{3} = 2\frac{1}{3}$ hours
18. $.2 \times 56.4 = 11.28$ gal
19. $432 \times .6 = 259.2$ mi
20. 4,000 meters
    approximately 4,000 yards

Lesson Practice 11A

1. done
2. done
3. $30\% = .30$
4. $85\% = .85$
5. $10\% = .1$
6. $9\% = .09$
7. done
8. $5\% = \frac{5}{100} = \frac{1}{20}$
9. $75\% = \frac{75}{100} = \frac{3}{4}$
10. $\frac{1}{1} = 1.00 = 100\%$
11. $\frac{1}{2} = .5 = 50\%$
12. $\frac{1}{4} = .25 = 25\%$
13. $\frac{3}{4} = .75 = 75\%$
14. $\frac{1}{5} = .2 = 20\%$
15. $\$3.00 \times .06 = \$0.18$
16. $\$7.60 \times .15 = \$1.14$
17. $\$7.60 \times .50 = \$3.80$ off
    $\$7.60 - \$3.80 = \$3.80$ sale price
18. $\$3.80 \times .05 = \$0.19$ tax
    $\$3.80 + \$0.19 = \$3.99$ total

Lesson Practice 11B

1. $38\% = .38$
2. $1\% = .01$
3. $95\% = .95$
4. $71\% = .71$
5. $15\% = .15$
6. $3\% = .03$
7. $10\% = \frac{10}{100} = \frac{1}{10}$
8. $23\% = \frac{23}{100}$
9. $60\% = \frac{60}{100} = \frac{3}{5}$
10. $\frac{1}{4} = .25 = 25\%$

LESSON PRACTICE 11B - SYSTEMATIC REVIEW 11D

11. $\frac{3}{4} = .75 = 75\%$
12. $\frac{1}{5} = .2 = 20\%$
13. $\frac{1}{1} = 1.00 = 100\%$
14. $\frac{1}{2} = .5 = 50\%$
15. $\$7.00 \times .05 = \$0.35$
16. $\$10.50 \times .16 = \$1.68$
17. $\$12.50 \times .20 = \$2.50$ off
    $\$12.50 - \$2.50 = \$10.00$ sale price
    $\$10.00 \times .09 = \$0.90$ tax
    $\$10.00 + \$0.90 = \$10.90$ total
18. $.25 \times 80 = 20$ happy people

Lesson Practice 11C

1. $47\% = .47$
2. $5\% = .05$
3. $69\% = .69$
4. $18\% = .18$
5. $32\% = .32$
6. $2\% = .02$
7. $15\% = \frac{15}{100} = \frac{3}{20}$
8. $17\% = \frac{17}{100}$
9. $50\% = \frac{50}{100} = \frac{1}{2}$
10. $\frac{1}{2} = .5 = 50\%$
11. $\frac{1}{5} = .2 = 20\%$
12. $\frac{3}{4} = .75 = 75\%$
13. $\frac{1}{4} = .25 = 25\%$
14. $\frac{1}{1} = 1.00 = 100\%$
15. done
16. $\$45.00 \times .30 = \$13.50$ off
    $\$45.00 - \$13.50 = \$31.50$ sale price
    $\$31.50 \times .08 = \$2.52$ tax
    $\$31.50 + \$2.52 = \$34.02$ total

17. $\$18.50 + \$9.50 = \$28.00$ total meals
    $\$28.00 \times .2 = \$5.60$ tip
    $\$28.00 + \$5.60 = \$33.60$ total meals and tip
18. $.50 \times 20 = 10$ people own sleds
    or $\frac{1}{2} \times 20 = 10$ people own sleds

Systematic Review 11D

1. done
2. $88\% = .88 = \frac{88}{100} = \frac{22}{25}$
3. $100\% = 1 = \frac{100}{100} = 1$
4. $25\% = .25 = \frac{25}{100} = \frac{1}{4}$
5. 50,000 cl
6. 1,200 dm
7. $200 \div 2 = 100$
8. $84 \div 4 = 21$
   $21 \times 3 = 63$
9. $99 \div 3 = 33$
10. 64
11. 1
12. 1,000
13. done
14. $5\frac{5}{8} \times 1\frac{2}{5} = \frac{\cancel{45}^9}{8} \times \frac{7}{\cancel{5}} = \frac{63}{8} = 7\frac{7}{8}$
15. $50 \times .96 = 48$ questions right
    $50 - 48 = 2$ questions wrong
16. $\$120.00 \times .25 = \$30.00$ off
    $\$120.00 - \$30.00 = \$90.00$ sale price
    $\$90.00 \times .05 = \$4.50$ tax
    $\$90.00 + \$4.50 = \$94.50$ total
17. 1,000 pieces
    approximately 1 yard
18. $2\frac{2}{3} \times 4\frac{1}{2} = \frac{\cancel{8}^4}{\cancel{3}} \times \frac{\cancel{9}^3}{\cancel{2}} = \frac{12}{1} = 12$ miles

Systematic Review 11E

1. $30\% = .3 = \dfrac{30}{100} = \dfrac{3}{10}$
2. $72\% = .72 = \dfrac{72}{100} = \dfrac{18}{25}$
3. $50\% = .5 = \dfrac{50}{100} = \dfrac{1}{2}$
4. $20\% = .2 = \dfrac{20}{100} = \dfrac{1}{5}$
5. 50 cl
6. 18,000 ml
7. $120 \div 3 = 40$
8. $72 \div 9 = 8$
   $8 \times 2 = 16$
9. $600 \div 6 = 100$
   $100 \times 5 = 500$
10. 2.448
11. .36
12. .31
13. 1.26
14. $4\dfrac{2}{7} \times 3\dfrac{1}{5} = \dfrac{\cancel{30}^{6}}{7} \times \dfrac{16}{\cancel{5}} = \dfrac{96}{7} = 13\dfrac{5}{7}$
15. $9\dfrac{1}{5} \times 5\dfrac{4}{5} = \dfrac{46}{5} \times \dfrac{29}{5} = \dfrac{1,334}{25} = 53\dfrac{9}{25}$
16. 45% of $20 = .45 \times 20 = 9$ boys
17. $\$85.00 \times .08 = \$6.80$
    shipping and handling
    $\$85.00 + \$6.80 = \$91.80$ total
    $\$91.80 - \$57.00 = \$34.80$ still needed
18. 100 dkm
19. $\$9.95 \times 5 = \$49.75$
20. $.52 \text{ lb} + .75 \text{ lb} = 1.27 \text{ lb}$
    $1.27 \times .5 = .635$ lb

Systematic Review 11F

1. $55\% = .55 = \dfrac{55}{100} = \dfrac{11}{20}$
2. $40\% = .4 = \dfrac{40}{100} = \dfrac{2}{5}$
3. $75\% = .75 = \dfrac{75}{100} = \dfrac{3}{4}$
4. $24\% = .24 = \dfrac{24}{100} = \dfrac{6}{25}$
5. 2,200 dg
6. 100 dkg
7. $6\dfrac{2}{3} + 1\dfrac{7}{9} = 6\dfrac{18}{27} + 1\dfrac{21}{27} =$
   $7\dfrac{39}{27} = 8\dfrac{12}{27} = 8\dfrac{4}{9}$
8. $7\dfrac{3}{10} + 9\dfrac{4}{5} = 7\dfrac{15}{50} + 9\dfrac{40}{50} =$
   $16\dfrac{55}{50} = 17\dfrac{5}{50} = 17\dfrac{1}{10}$
9. $3\dfrac{2}{7} - 1\dfrac{2}{3} = 3\dfrac{6}{21} - 1\dfrac{14}{21} =$
   $2\dfrac{27}{21} - 1\dfrac{14}{21} = 1\dfrac{13}{21}$
10. .8547
11. 4.32
12. 1.77
13. 3.33
14. $1\dfrac{2}{3} \times 4\dfrac{3}{5} = \dfrac{\cancel{5}}{3} \times \dfrac{23}{\cancel{5}} = \dfrac{23}{3} = 7\dfrac{2}{3}$
15. $7\dfrac{3}{5} \times 1\dfrac{1}{2} = \dfrac{\cancel{38}^{19}}{5} \times \dfrac{3}{\cancel{2}} = \dfrac{57}{5} = 11\dfrac{2}{5}$
16. 85% of $20 = .85 \times 20 = 17$ games won
17. $\$1.99 \times .15 = \$0.2985$
    (rounded to $0.30) off
    $\$1.99 - \$0.30 = \$1.69$ for one fish
    $\$1.69 \times 3 = \$5.07$ for three fish
18. $40 \times .6 = 24$ mi
19. $8 \times 1.06 = 8.48$ qt
20. $6\dfrac{2}{3} \times \dfrac{3}{5} = \dfrac{\cancel{20}^{4}}{\cancel{3}} \times \dfrac{\cancel{3}}{\cancel{5}} = \dfrac{4}{1} =$
    $\dfrac{4}{1} = 4$ bales spoiled

Lesson Practice 12A

1. done
2. $9 = \dfrac{900}{100} = 900\%$
3. $5 = \dfrac{500}{100} = 500\%$
4. done

LESSON PRACTICE 12A - LESSON PRACTICE 12C

5. $\frac{1}{4} = \frac{25}{100} = 25\%$

6. $\frac{3}{4} = \frac{75}{100} = 75\%$

7. $\frac{1}{5} = \frac{20}{100} = 20\%$

8. $\frac{2}{5} = \frac{40}{100} = 40\%$

9. $\frac{3}{5} = \frac{60}{100} = 60\%$

10. $1\frac{1}{4} = \frac{100}{100} + \frac{25}{100} = \frac{125}{100} = 125\%$

11. $2\frac{1}{2} = \frac{200}{100} + \frac{50}{100} = \frac{250}{100} = 250\%$

12. $3\frac{3}{4} = \frac{300}{100} + \frac{75}{100} = \frac{375}{100} = 375\%$

13. $5\frac{1}{5} = \frac{500}{100} + \frac{20}{100} = \frac{520}{100} = 520\%$

14. done

15. $250\% = 2.5$

16. $520\% = 5.2$

17. $100\% + 15\% + 6\% = 121\% = 1.21$
    $1.21 \times \$18.90 = \$22.869$
    (rounds to $22.87)

18. $400\% = 4$
    $4 \times \$0.45 = \$1.80$

Lesson Practice 12B

1. $8 = \frac{800}{100} = 800\%$

2. $6 = \frac{600}{100} = 600\%$

3. $2 = \frac{200}{100} = 200\%$

4. $\frac{2}{5} = \frac{40}{100} = 40\%$

5. $\frac{1}{5} = \frac{20}{100} = 20\%$

6. $\frac{4}{5} = \frac{80}{100} = 80\%$

7. $\frac{1}{10} = \frac{10}{100} = 10\%$

8. $\frac{9}{10} = \frac{90}{100} = 90\%$

9. $\frac{1}{4} = \frac{25}{100} = 25\%$

10. $3\frac{1}{10} = \frac{300}{100} + \frac{10}{100} = \frac{310}{100} = 310\%$

11. $6\frac{3}{4} = \frac{600}{100} + \frac{75}{100} = \frac{675}{100} = 675\%$

12. $5\frac{1}{2} = \frac{500}{100} + \frac{50}{100} = \frac{550}{100} = 550\%$

13. $2\frac{3}{5} = \frac{200}{100} + \frac{60}{100} = \frac{260}{100} = 260\%$

14. $310\% = 3.10$

15. $675\% = 6.75$

16. $100\% = 1$

17. $100\% + 20\% + 5\% = 125\% = 1.25$
    $1.25 \times \$30.00 = \$37.50$

18. $100\% + 5\% = 105\% = 1.05$
    $1.05 \times \$32.95 = \$34.5975$
    (rounds to $34.60)

Lesson Practice 12C

1. $3 = \frac{300}{100} = 300\%$

2. $7 = \frac{700}{100} = 700\%$

3. $1 = \frac{100}{100} = 100\%$

4. $\frac{1}{2} = \frac{50}{100} = 50\%$

5. $\frac{3}{4} = \frac{75}{100} = 75\%$

6. $\frac{3}{5} = \frac{60}{100} = 60\%$

7. $\frac{5}{10} = \frac{50}{100} = 50\%$

8. $\frac{1}{25} = \frac{4}{100} = 4\%$

9. $\frac{1}{50} = \frac{2}{100} = 2\%$

10. $7\frac{1}{4} = \frac{700}{100} + \frac{25}{100} = \frac{725}{100} = 725\%$

11. $4\frac{1}{5} = \frac{400}{100} + \frac{20}{100} = \frac{420}{100} = 420\%$

12. $1\frac{3}{4} = \frac{100}{100} + \frac{75}{100} = \frac{175}{100} = 175\%$

13. $3\frac{5}{10} = \frac{300}{100} + \frac{50}{100} = \frac{350}{100} = 350\%$

14. $725\% = 7.25$

15. $420\% = 4.2$

LESSON PRACTICE 12C - SYSTEMATIC REVIEW 12F

16. 850% = 8.5
17. 100% + 18% + 2% = 120% = 1.2
    $1.2 \times \$55.20 = \$66.24$
18. 100% = 1
    $1 \times 20 = 20$

Systematic Review 12D

1. done
2. $6\frac{3}{4} = \frac{600}{100} + \frac{75}{100} = \frac{675}{100} = 6.75 = 675\%$
3. $25\% = .25 = \frac{25}{100} = \frac{1}{4}$
4. $50\% = .5 = \frac{50}{100} = \frac{1}{2}$
5. 600 m
6. 7,000 mm
7. 1,000
8. 100
9. 10
10. $\frac{1}{10}$
11. $\frac{1}{100}$
12. $\frac{1}{1,000}$
13. $\frac{3}{4} \div \frac{1}{4} = \frac{3 \div 1}{1} = 3$
14. $\frac{4}{5} \div \frac{1}{3} = \frac{12}{15} \div \frac{5}{15} = \frac{12 \div 5}{1} = 2\frac{2}{5}$
15. $\frac{2}{3} \div \frac{1}{4} = \frac{8}{12} \div \frac{3}{12} = \frac{8 \div 3}{1} = \frac{8}{3} = 2\frac{2}{3}$
16. $100\% + 6\% + 5\% = 111\% = 1.11$
    $1.11 \times \$36.00 = \$39.96$
17. $100\% - 25\% = 75\% = .75$
    $.75 \times \$39.96 = \$29.97$
18. Yes; $(10\% = .1 = \frac{1}{10})$
    $.1 \times 100 = 10$

Systematic Review 12E

1. $2\frac{1}{4} = \frac{200}{100} + \frac{25}{100} = \frac{225}{100} =$
   $225\% = 2.25$
2. $8\frac{1}{5} = \frac{800}{100} + \frac{20}{100} = \frac{820}{100} =$
   $820\% = 8.20$
3. $42\% = .42 = \frac{42}{100} = \frac{21}{50}$
4. $150\% = 1.5 = \frac{150}{100} = 1\frac{50}{100} = 1\frac{1}{2}$
5. yard
6. inches
7. quarts
8. ounce
9. pounds
10. miles
11. inch
12. $\frac{7}{8} \div \frac{1}{3} = \frac{21}{24} \div \frac{8}{24} = \frac{21 \div 8}{1} =$
    $\frac{21}{8} = 2\frac{5}{8}$
13. $\frac{5}{6} \div \frac{1}{2} = \frac{10}{12} \div \frac{6}{12} = \frac{10 \div 6}{1} =$
    $\frac{10}{6} = 1\frac{4}{6} = 1\frac{2}{3}$
14. $\frac{2}{3} \div \frac{1}{4} = \frac{8}{12} \div \frac{3}{12} = \frac{8 \div 3}{1} =$
    $\frac{8}{3} = 2\frac{2}{3}$
15. $\frac{5}{8} \div \frac{1}{16} = \frac{80}{128} \div \frac{8}{128} =$
    $\frac{80 \div 8}{1} = \frac{10}{1} = 10$ pies
16. $\$235.00 \times .05 = \$11.75$ tax
    $\$235.00 + \$11.75 = \$246.75$
17. $100\% - 40\% = 60\% = .6$
    $.6 \times \$12.4 = \$7.44$
18. $68 \times 2.5 = 170$ cm
19. $170 \times 10 = 1,700$ mm
20. Yes; $(50\% = \frac{1}{2})$
    $\frac{1}{2} \times \frac{\$40.00}{1} = \frac{\$40.00}{2} = \$20.00$
    or
    $.5 \times \$40.00 = \$20.00$

Systematic Review 12F

1. $7\frac{3}{4} = \frac{700}{100} + \frac{75}{100} = \frac{775}{100} = 775\% = 7.75$
2. $5\frac{1}{10} = \frac{500}{100} + \frac{10}{100} = \frac{510}{100} = 510\% = 5.10$
3. $9\% = .09 = \frac{9}{100}$
4. $100\% = 1 = \frac{100}{100} = 1$
5. 3 feet or 1 yard
6. .4 inch
7. 2.5 cm
8. 28 grams
9. 2.2 lb
10. .6 miles
11. 1.06 qt
12. $\frac{6}{7} \div \frac{4}{7} = \frac{6 \div 4}{1} = \frac{6}{4} = 1\frac{2}{4} = 1\frac{1}{2}$
13. $\frac{3}{5} \div \frac{1}{3} = \frac{9}{15} \div \frac{5}{15} = \frac{9 \div 5}{1} = \frac{9}{5} = 1\frac{4}{5}$
14. $\frac{5}{8} \div \frac{2}{3} = \frac{15}{24} \div \frac{16}{24} = \frac{15 \div 16}{1} = \frac{15}{16}$
15. $\frac{4}{5} \div \frac{2}{5} = \frac{4 \div 2}{1} = \frac{4}{2} = 2$ friends
16. $20 \times 3 = 60$ balloons
17. $100\% + 20\% + 4\% = 124\% = 1.24$
    $1.24 \times \$15.00 = \$18.60$
18. $20 \times 28 = 560$ g
19. $560 \times 1,000 = 560,000$ mg
20. $100\% - 20\% = 80\% = .8$
    $.8 \times \$355.00 = \$284.00$

Lesson Practice 13A

1. brown
2. $.5 \times 20 = 10$
3. 5 (There are the same number of blondes as redheads.)
4. vanilla
5. strawberry
6. $300 \times .3 = 90$

7.

 grew
  did not grow

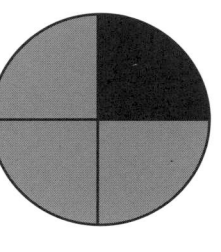

(The position of the colored sections is not important as long as the right number of sections are colored.)

8.

  giving
  savings
  other

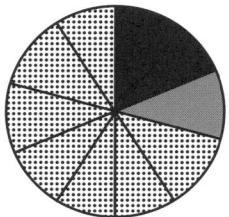

Lesson Practice 13B

1. sleep
2. $.25 \times 24 = 6$
3. $25\% = \frac{25}{100} = \frac{1}{4}$
4. repairs
5. gasoline
6. $\$1,000.00 \times .3 = \$300.00$

7.

  cats only
  dogs only
 both

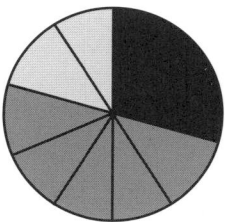

8.

 ice cream
  pie
  cake

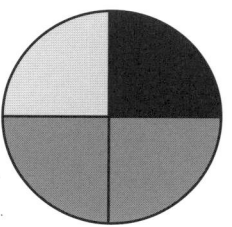

Lesson Practice 13C

1. red, white, and blue
2. blue and white
3. $600 \times .35 = 210$
4. pine
5. $480 \times .4 = 192$
6. $480 \times .3 = 144$
7.

 goldfish

 minnows

 guppies

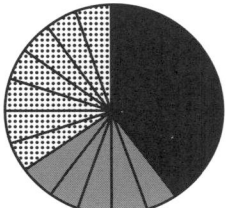

8.

 wood

 brick

 stone

12. $1.79 \times .3 = .537$
13. $25 - .45 = 24.55$
14. $.135 + .4 = .535$
15. $4\frac{2}{3} \div 1\frac{1}{5} = \frac{14}{3} \div \frac{6}{5} = \frac{70}{15} \div \frac{18}{15} =$
    $\frac{70 \div 18}{1} = \frac{70}{18} = \frac{35}{9} = 3\frac{8}{9}$
16. $4\frac{1}{8} \div 2\frac{2}{3} = \frac{33}{8} \div \frac{8}{3} = \frac{99}{24} \div \frac{64}{24} =$
    $\frac{99 \div 64}{1} = \frac{99}{64} = 1\frac{35}{64}$
17. $7\frac{5}{10} \div 1\frac{1}{2} = \frac{75}{10} \div \frac{3}{2} = \frac{150}{20} \div \frac{30}{20} =$
    $\frac{150 \div 30}{1} = \frac{150}{30} = 5$ bags
18. $100\% + 7\% = 107\% = 1.07$
    $\$35.00 \times 1.07 = \$37.45$

Systematic Review 13D

1. corn
2. $10\% + 15\% = 25\% = \frac{25}{100} = \frac{1}{4}$
3. $500 \times .75 = 375$
4. $25\% = .25 = \frac{25}{100} = \frac{1}{4}$
5. $40\% = .4 = \frac{40}{100} = \frac{2}{5}$
6. $300\% = 3 = \frac{300}{100} = 3$
7. $15\% = .15 = \frac{15}{100} = \frac{3}{20}$
8. $4\frac{1}{2} = \frac{400}{100} + \frac{50}{100} = \frac{450}{100} =$
   $450\% = 4.5$
9. 2.2 lb
10. .6 miles
11. 1.06 qt

Systematic Review 13E

1.

 biking

 running

 swimming

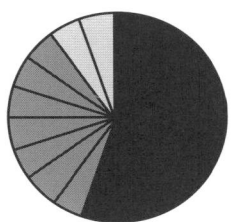

2. $50 \text{ miles} \times .1 = 5$ miles
3. $50\% = .5 = \frac{50}{100} = \frac{1}{2}$
4. $20\% = .2 = \frac{20}{100} = \frac{1}{5}$
5. $5\% = .05 = \frac{5}{100} = \frac{1}{20}$
6. $700\% = 7 = \frac{700}{100} = 7$
7. $6\frac{1}{4} = \frac{600}{100} + \frac{25}{100} =$
   $\frac{625}{100} = 625\% = 6.25$
8. .4 inches
9. 2.5 cm
10. 28 grams
11. $2.45 \times .09 = .2205$

SYSTEMATIC REVIEW 13E - LESSON PRACTICE 14B

12. $.7-.15=.55$
13. $1.58+7.6=9.18$
14. $3\frac{1}{6} \div 1\frac{1}{5} = \frac{19}{6} \div \frac{6}{5} = \frac{95}{30} \div \frac{36}{30} =$
    $\frac{95 \div 36}{1} = \frac{95}{36} = 2\frac{23}{36}$
15. $1\frac{1}{4} \div 1\frac{1}{8} = \frac{5}{4} \div \frac{9}{8} = \frac{40}{32} \div \frac{36}{32} =$
    $\frac{40 \div 36}{1} = \frac{40}{36} = \frac{10}{9} = 1\frac{1}{9}$
16. $2\frac{1}{6} \div 1\frac{4}{5} = \frac{13}{6} \div \frac{9}{5} = \frac{65}{30} \div \frac{54}{30} =$
    $\frac{65 \div 54}{1} = \frac{65}{54} = 1\frac{11}{54}$
17. $6\frac{3}{4} \div 1\frac{1}{4} = \frac{27}{4} \div \frac{5}{4} = \frac{27 \div 5}{1} =$
    $\frac{27}{5} = 5\frac{2}{5}$
    5 people can be served. There will pizza left over because she can't serve pizza to a part of a person.
18. $100\% + 10\% = 110\% = 1.1$
    $\$1.30 \times 1.1 = \$1.43$
19. $55.6 \times 7 = 389.2$ miles
20. $2 \times 2.2 = 4.4$ lb

Systematic Review 13F

1. clouds
2. $30 \times .2 = 6$ days
3. $30 \times .1 = 3$ days
4. $75\% = .75 = \frac{75}{100} = \frac{3}{4}$
5. $2\% = .02 = \frac{2}{100} = \frac{1}{50}$
6. $500\% = 5 = \frac{500}{100} = 5$
7. $30\% = .3 = \frac{30}{100} = \frac{3}{10}$
8. $3\frac{1}{5} = \frac{300}{100} + \frac{20}{100} = \frac{320}{100} = 320\% = 3.2$
9. 6,000 g
10. 25,000 ml
11. $.17 \times .58 = .0986$
12. $1.75 - .9 = .85$
13. $3.68 + .061 = 3.741$

14. $3\frac{1}{2} \div 2\frac{2}{5} = \frac{7}{2} \div \frac{12}{5} = \frac{35}{10} \div \frac{24}{10} =$
    $\frac{35 \div 24}{1} = \frac{35}{24} = 1\frac{11}{24}$
15. $5\frac{1}{2} \div 2\frac{2}{3} = \frac{11}{2} \div \frac{8}{3} = \frac{33}{6} \div \frac{16}{6} =$
    $\frac{33 \div 16}{1} = \frac{33}{16} = 2\frac{1}{16}$
16. $3\frac{1}{5} \div 1\frac{3}{4} = \frac{16}{5} \div \frac{7}{4} = \frac{64}{20} \div \frac{35}{20} =$
    $\frac{64 \div 35}{1} = \frac{64}{35} = 1\frac{29}{35}$
17. 200 cm
    $200 \times .4 = 80$ in
18. $60 \times 1.2 = 72$
19. $1.8 \times 8 = 14.4$ cups
20. $50 \times .6 = 30$ mi

Lesson Practice 14A

1. done
2. done
3. Estimate:
   ```
      20.
   ×   .1
      2.0
   ```

   ```
       22.163           22 163   3 places
   ×     .11          ×     11   2 places
       .22163            22 163
      2.2163            2 2163
      2.43793           2.43793  5 places
   ```
4. Estimate:
   ```
     8
   × 2
    16
   ```

   ```
       8.246            8 246    3 places
   ×   2.159        ×   2 159    3 places
       .074214            742 14
       .41230            4 1230
       .8246             8 246
     16.492            16 492
     17.803114         17.803 114  6 places
   ```

5. Estimate:

    100.
    ×   .6
    60.

    135.004
    ×   .58
    10.80032
    67.5020
    78.30232

    1 3 5 0 0 4   3 places
    ×   1 1   2 places
    10 80032
    67 5020
    78.30232   5 places

6. Estimate:

    40.
    ×   .6
    24.

    41.685
    ×   .555
    .208425
    2.08425
    20.8425
    23.135175

    4 1 6 8 5   3 places
    ×   5 5 5   3 places
    208425
    208425
    208425
    23.135175   6 places

7. 9.625 × .25 = 2.40625 mi
8. 12.75 × $1.539 = $19.62225
   (rounded to $19.62)

Lesson Practice 14B

1. Estimate:

    50.
    ×   .005
    .25

    50.345
    ×   .005
    .251725

    50345   3 places
    ×   005   3 places
    .251725   6 places

2. Estimate:

    31
    ×   2
    62

    31.489
    ×   2.23
    .94467
    6.2978
    62.978
    70.22047

    3 1489   3 places
    ×   223   2 places
    94467
    6 2978
    62 978
    70.22047   5 places

3. Estimate:

    .9
    ×.009
    .0081

    .876
    ×.009
    .007884

    876   3 places
    ×   9   3 places
    .007884   6 places

4. Estimate :

    6
    ×1
    6

    6.218
    ×1.064
    .024872
    .37308
    6.218
    6.615952

    6218   3 places
    ×   1064   3 places
    24872
    37308
    6 218
    6.615952   6 places

5. Estimate:

    500.
    ×   .2
    100

    450.132
    ×   .222
    .900264
    9.00264
    90.0264
    99.929304

    450 132   3 places
    ×   222   3 places
    900264
    9 00264
    90 0264
    99.929304   6 places

6. Estimate:

    2
    × .8
    1.6

    1.539
    × .82
    .03078
    1.2312
    1.26198

    1539   3 places
    ×   82   2 places
    3078
    1 2312
    1.26198   5 places

7. 39.37 × .001 = .03937 in
8. $1.279 × 1.30 = $1.6627
   (rounded to $1.663)

Lesson Practice 14C

1. Estimate:

   60
   × .03
   ―――
   1.80

   61.198
   × .03
   ―――
   1.83594

   6 1198   3 places
   ×   3   2 places
   ―――――――
   1.83594   5 places

2. Estimate:

   40
   × 1
   ――
   40

   42.345
   × 1.16
   ―――――
   2.54070
   4.2345
   42.345
   ―――――
   49.12020

   42345   3 places
   ×   116   2 places
   ――――――
   2 54070
   4 2345
   42 345
   ――――――
   49.12020   5 places

3. Estimate:

   .5
   ×.8
   ――
   .40

   .511
   ×.8
   ――
   .4088

   511   3 places
   ×   8   1 place
   ―――――
   .4088   4 places

4. Estimate:

   8
   × 3
   ――
   24

   7.639
   × 2.804
   ―――――
   .030556
   6.1112
   15.278
   ―――――
   21.419756

   7639   3 places
   × 2804   3 places
   ―――――――
   30556
   6 11120
   15 278
   ―――――――
   21.419756   6 places

5. Estimate:

   20
   × .1
   ――
   2

   20.352
   × .136
   ―――――
   .122112
   .61056
   2.0352
   ―――――
   2.767872

   20352   3 places
   ×   136   3 places
   ――――――
   122112
   61056
   2 0352
   ――――――
   2.767872   6 places

6. Estimate:

   9
   × .7
   ――
   6.3

   9.293
   × .71
   ―――――
   .09293
   6.5051
   ―――――
   6.59803

   9293   3 places
   ×   71   2 places
   ――――――
   9293
   6 5051
   ――――――
   6.59803   5 places

   9293   3 places
   x71   +2 places
   ――――
   9293
   65051
   ――――
   6.59803   5 places

7. $1.06 \times .001 = .00106$ qt
8. $1.667 \times 60 = 100.02$ gal

Systematic Review 14D

1. Estimate:

   2
   × .6
   ――
   1.2

   1724   3 places
   ×   6   1 place
   ――――――
   1.0344   4 places

2. Estimate:

   3
   × .4
   ――
   1.2

   3465   3 places
   ×   403   3 places
   ――――――
   10395
   1 3860
   ――――――
   1.396395   6 places

3. $50\% = .5 = \dfrac{50}{100} = \dfrac{1}{2}$

4. $8\% = .08 = \dfrac{8}{100} = \dfrac{2}{25}$

5. $600\% = 6 = \dfrac{600}{100} = 6$

6. $20\% = .2 = \dfrac{20}{100} = \dfrac{1}{5}$

7. $5\dfrac{1}{4} = \dfrac{500}{100} + \dfrac{25}{100} = \dfrac{525}{100} = 525\% = 5.25$

8. 70,000 cm

9. 100 mg

10. $6\dfrac{1}{8} + 3\dfrac{1}{2} = 6\dfrac{2}{16} + 3\dfrac{8}{16} = 9\dfrac{10}{16} = 9\dfrac{5}{8}$

11. $10\dfrac{2}{5} - 4\dfrac{5}{8} = 10\dfrac{16}{40} - 4\dfrac{25}{40} =$
    $9\dfrac{56}{40} - 4\dfrac{25}{40} = 5\dfrac{31}{40}$

12. $\dfrac{\cancel{7}}{3} \times \dfrac{\cancel{6}^2}{\cancel{7}} = \dfrac{2}{1} = 2$

13. $\dfrac{3}{5} \times \dfrac{2}{1} = \dfrac{6}{5} = 1\dfrac{1}{5}$

14. $\dfrac{7}{\cancel{2}8} \times \dfrac{\cancel{4}}{3} = \dfrac{7}{6} = 1\dfrac{1}{6}$

15. $\dfrac{3}{8} \times \dfrac{5}{1} = \dfrac{15}{8} = 1\dfrac{7}{8}$

16. $\dfrac{10}{16} \div \dfrac{1}{8} = \dfrac{10}{\cancel{16}_2} \times \dfrac{\cancel{8}}{1} = \dfrac{10}{2} = 5$ pieces

17. $\$15.50 \times 1.06 = \$16.43$

18. $.6 \times .001 = .0006$ mi

6. $25\% = .25 = \dfrac{25}{100} = \dfrac{1}{4}$

7. $100\% = 1 = \dfrac{100}{100} = 1$

8. $250\% = 2.5 = \dfrac{250}{100} = 2\dfrac{1}{2}$

9. $16\% = .16 = \dfrac{16}{100} = \dfrac{4}{25}$

10. $1\dfrac{3}{4} = \dfrac{100}{100} + \dfrac{75}{100} = \dfrac{175}{100} = 175\% = 1.75$

11. 12,000 L

12. 300 cm

13. $\dfrac{5}{\cancel{3}9} \times \dfrac{\cancel{3}}{2} = \dfrac{5}{6}$

14. $\dfrac{9}{\cancel{2}10} \times \dfrac{\cancel{5}}{1} = \dfrac{9}{2} = 4\dfrac{1}{2}$

15. $\dfrac{1}{3} \times \dfrac{4}{3} = \dfrac{4}{9}$

16. $8 \times .08 = .64$ hours

17. $15.25 \times \$1.549 =$
    $\$23.62225$ (rounded $\$23.62$)

18. $1.56 + 2.13 + 1.5 + .66 = 5.85$ lb

19. $5.85 \times 16 = 93.6$ oz
    $93.6 \times 28 = 2{,}620.8$ g

20. $.25 \times 12 = 3$ oranges to her mother
    $\dfrac{1}{4} \times \dfrac{12}{1} = \dfrac{12}{4} = 3$ oranges to her sister
    $12 - 6 = 6$ left
    $6 \div 12 = .5 = 50\%$ left

Systematic Review 14E

1. Estimate:

    ```
          2              2 1 9 2    3 places
        × .007         ×       7    3 places
        ──────         ─────────
         .014           .015344    6 places
    ```

2. Estimate:

    ```
         10            1 3 5 4 3    3 places
        × .8         ×        75    2 places
        ────         ─────────
          8             6 7 7 1 5
                        9 4 8 0 1
                      ─────────────
                       10.15725    5 places
    ```

3. summer
4. spring
5. $850 \times .10 = 85$

Systematic Review 14F

1. Estimate:

    ```
         .6                567    3 places
        ×.1              ×  148    3 places
        ────             ──────
         .06               4536
                           2268
                            567
                         ─────────
                         .083916   6 places
    ```

SYSTEMATIC REVIEW 14F - LESSON PRACTICE 15C

2. Estimate:

    2
    $\underline{\times 1}$
    2

    $1\,9\,0\,5$ 3 places
    $\underline{\times 1\,3\,2\,1}$ 3 places
    $1\,9\,0\,5$
    $3\,8\,1\,0$
    $5\,7\,1\,5$
    $\underline{1\,9\,0\,5}$
    $2.5\,1\,6\,5\,0\,5$ 6 places

3. oak
    maple
    hemlock
   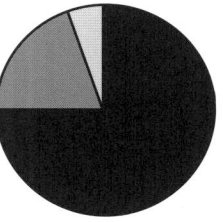

4. $600 \times .05 = 30$ trees are hemlock
5. $75\% = .75 = \frac{75}{100} = \frac{3}{4}$
6. $45\% = .45 = \frac{45}{100} = \frac{9}{20}$
7. $125\% = 1.25 = \frac{125}{100} = 1\frac{1}{4}$
8. $1\% = .01 = \frac{1}{100}$
9. $7\frac{1}{5} = \frac{700}{100} + \frac{20}{100} = \frac{720}{100} = 720\% = 7.2$
10. 35,000 mg
11. 1,100 dkm
12. 500 cm
13. $\frac{2}{5} \times \frac{8}{3} = \frac{16}{15} = 1\frac{1}{15}$
14. $\frac{\cancel{4}^2}{5} \times \frac{3}{\cancel{2}} = \frac{6}{5} = 1\frac{1}{5}$
15. $\frac{\cancel{3}}{\cancel{4}} \times \frac{\cancel{4}}{\cancel{3}} = 1$
16. $900 \times .15 = 135$ yellow roses
17. $\$9.00 \times 1.10 = \$9.90$
18. $90 \times .4 = 36$ in
19. $5.725 \times 0.1 = .5725$ tons
20. $.5725 \times 2,000 = 1,145$ lb

Lesson Practice 15A
1. done
2. done
3. 7.5 dkg
4. .014 km
5. .4 km
6. 5.3 L
7. 1.3 cg
8. .48 dkm
9. .07 hl
10. .000609 kg
11. .5 cm
12. 1.2 dkl

Lesson Practice 15B
1. 30 L
2. 5 m
3. 10 kl
4. .25 km
5. .18 g
6. .45 kl
7. .5 km
8. 80 g
9. 8.3 cg
10. .008 kl
11. 1.02 m
12. .01 km

Lesson Practice 15C
1. .015 dkm
2. 8 g
3. 1 hm
4. .25 km
5. 30 g
6. .034 dkg
7. .99 m
8. 7.3 dkm
9. 15 kl
10. 50 m
11. 45.638 g
12. 300 kg

Systematic Review 15D

1. .45 g
2. 60 cm
3. 100 mg
4. Estimate:

   $$\begin{array}{r} .6 \\ \times .5 \\ \hline .30 \end{array}$$

   $$\begin{array}{r} 621 \quad \text{3 places} \\ \times\ 543 \quad \text{3 places} \\ \hline 1863 \\ 2484 \\ 3105 \\ \hline .337203 \quad \text{6 places} \end{array}$$

5. Estimate:

   $$\begin{array}{r} 3 \\ \times 2 \\ \hline 6 \end{array}$$

   $$\begin{array}{r} 2728 \quad \text{3 places} \\ \times\ 169 \quad \text{2 places} \\ \hline 24552 \\ 16368 \\ 2728 \\ \hline 4.61032 \quad \text{5 places} \end{array}$$

6. $11\% = .11 = \frac{11}{100}$
7. $5\% = .05 = \frac{5}{100} = \frac{1}{20}$
8. $100\% = 1 = \frac{100}{100} = 1$
9. $25\% = .25 = \frac{25}{100} = \frac{1}{4}$
10. $18 \div 2 = 9$
    $9 \times 1 = 9$
11. $69 \div 3 = 23$
    $23 \times 2 = 46$
12. $248 \div 8 = 31$
    $31 \times 5 = 155$
13. $4 \div 4 = 1$
14. $7\frac{7}{8} \div 1\frac{1}{2} = \frac{63}{8} \div \frac{3}{2} =$

    $\frac{\cancel{63}^{21}}{\cancel{8}_4} \times \frac{\cancel{2}}{\cancel{3}} = \frac{21}{4} = 5\frac{1}{4}$
15. $3\frac{2}{7} \div 2\frac{1}{3} = \frac{23}{7} \div \frac{7}{3} =$

    $\frac{23}{7} \times \frac{3}{7} = \frac{69}{49} = 1\frac{20}{49}$
16. $24 \times .1 = 2.4$ dkm
17. $105 \times 2.2 = 231$ lb
18. $\$1.679 \times 2 = \$3.358$

Systematic Review 15E

1. .39 m
2. 3,800 g
3. .087 kl
4. Estimate:

   $$\begin{array}{r} 50 \\ \times\ .6 \\ \hline 30 \end{array}$$

   $$\begin{array}{r} 48003 \quad \text{3 places} \\ \times\ \ \ \ \ 61 \quad \text{2 places} \\ \hline 48003 \\ 288018 \\ \hline 29.28183 \quad \text{5 places} \end{array}$$

5. Estimate:

   $$\begin{array}{r} 20 \\ \times\ 30 \\ \hline 600 \end{array}$$

   $$\begin{array}{r} 2277 \quad \text{2 places} \\ \times\ 3154 \quad \text{2 places} \\ \hline 9108 \\ 11385 \\ 2277 \\ 6831 \\ \hline 718.1658 \quad \text{4 places} \end{array}$$

6. $250\% = 2.5 = \frac{250}{100} = 2\frac{50}{100} = 2\frac{1}{2}$
7. $34\% = .34 = \frac{34}{100} = \frac{17}{50}$
8. $7\% = .07 = \frac{7}{100}$
9. $75\% = .75 = \frac{75}{100} = \frac{3}{4}$
10. $400 \div 10 = 40$
    $40 \times 1 = 40$
11. $192 \div 6 = 32$
    $32 \times 5 = 160$
12. $96 \div 4 = 24$
    $24 \times 3 = 72$
13. $100 \div 5 = 20$
    $20 \times 4 = 80$
14. $6\frac{1}{3} \div 2\frac{1}{8} = \frac{19}{3} \div \frac{17}{8} =$

    $\frac{19}{3} \times \frac{8}{17} = \frac{152}{51} = 2\frac{50}{51}$
15. $8\frac{8}{10} \div 2\frac{2}{3} = \frac{88}{10} \div \frac{8}{3} =$

    $\frac{\cancel{88}^{11}}{10} \times \frac{3}{\cancel{8}} = \frac{33}{10} = 3\frac{3}{10}$

SYSTEMATIC REVIEW 15E - LESSON PRACTICE 16A

16. 61 km
17. 110 mg
18. $79.97 × .37 = $29.5889 off
    (rounded to $29.59)
    $79.97 − $29.59 = $50.38 sale price
19. $8.95 + $8.95 + $12.49 = $30.39
    $30.39 × 1.16 = $35.2524 total
    (rounded to $35.25)
20. $2\frac{2}{5} \times 1\frac{2}{3} = \frac{\cancel{12}^4}{\cancel{5}} \times \frac{\cancel{5}}{\cancel{3}} = 4$ mi

Systematic Review 15F
1. .005 g
2. 170,000 cm
3. .5 hl
4. Estimate:

   ```
        3              2952    3 places
   ×  .006           ×    6    3 places
   ─────             ─────────
     .018             .017712  6 places
   ```

5. Estimate:

   ```
      100              135123   3 places
   ×   .6           ×    .612   3 places
   ─────            ─────────
       60              270246
                      1 35123
                     81 0738
                    ─────────
                    82.695276  6 places
   ```

6. $2\% = .02 = \frac{2}{100} = \frac{1}{50}$
7. $20\% = .2 = \frac{20}{100} = \frac{1}{5}$
8. $60\% = .6 = \frac{60}{100} = \frac{3}{5}$
9. $900\% = 9 = \frac{900}{100} = 9$
10. 54 ÷ 6 = 9
    9 × 1 = 9
11. 88 ÷ 8 = 11
    11 × 3 = 33
12. 490 ÷ 7 = 70
    70 × 5 = 350

13. 300 ÷ 5 = 60
    60 × 2 = 120
14. $5\frac{1}{9} \div 2\frac{1}{6} = \frac{46}{9} \div \frac{13}{6} =$
    $\frac{46}{\cancel{9}_3} \times \frac{\cancel{6}^2}{13} = \frac{92}{39} = 2\frac{14}{39}$
15. $7\frac{1}{3} \div \frac{2}{5} = \frac{22}{3} \div \frac{2}{5} =$
    $\frac{\cancel{22}^{11}}{3} \times \frac{5}{\cancel{2}} = \frac{55}{3} = 18\frac{1}{3}$
16. $8\frac{3}{4} \div 5 = \frac{35}{4} \div \frac{5}{1} =$
    $\frac{\cancel{35}^7}{4} \times \frac{1}{\cancel{5}} = \frac{7}{4} = 1\frac{3}{4}$ lb
17. .5 kl
18. 68 × 2.5 = 170 cm = 1,700 mm
19. $48.00 × 1.25 = $60.00 from Dad
    $48.00 + $60.00 = $108.00 total
20. .2 × 100 = 20 days
    $\frac{1}{5} \times \frac{100}{1} = \frac{100}{5} = 20$ days

Lesson Practice 16A
1. done
2. done
3. A = 3.14(1.6)² = 8.0384 m²
   C = 2(3.14)(1.6) = 10.048 m
4. A = 3.14(15)² = 706.5 ft²
   C = (3.14)(30) = 94.2 ft
5. A = 3.14(12)² = 452.16 ft²
6. C = 2(3.14)(7.5) = 47.1 ft
7. C = (3.14)(12) = 37.68 in
   37.68" > 36"; it will not be enough
8. A = 3.14(9)² = 254.34 m²

Lesson Practice 16B

1. $A = 3.14(3)^2 = 28.26 \text{ in}^2$
   $C = 2(3.14)(3) = 18.84 \text{ in}$
2. $A = 3.14(5)^2 = 78.5 \text{ ft}^2$
   $C = (3.14)(10) = 31.4 \text{ ft}$
3. $A = 3.14(4.2)^2 = 55.3896 \text{ m}^2$
   $C = 2(3.14)(4.2) = 26.376 \text{ m}$
4. $A = 3.14(6)^2 = 113.04 \text{ ft}^2$
   $C = (3.14)(12) = 37.68 \text{ ft}$
5. $A = 3.14(16)^2 = 803.84 \text{ ft}^2$
6. $C = 2(3.14)(11) = 69.08 \text{ ft}$
7. $C = (3.14)(5.4) = 16.956 \text{ mm}$
8. $A = 3.14(8.5)^2 = 226.865 \text{ ft}^2$

Lesson Practice 16C

1. $A = 3.14(6)^2 = 113.04 \text{ in}^2$
   $C = 2(3.14)(6) = 37.68 \text{ in}$
2. $A = 3.14(7)^2 = 153.86 \text{ ft}^2$
   $C = (3.14)(14) = 43.96 \text{ ft}$
3. $A = 3.14(10)^2 = 314 \text{ m}^2$
   $C = 2(3.14)(10) = 62.8 \text{ m}$
4. $A = 3.14(4.75)^2 = 70.84625 \text{ ft}^2$
   $C = (3.14)(9.5) = 29.83 \text{ ft}$
5. $A = 3.14(6)^2 = 113.04 \text{ in}^2$
   (rounded to 113 in^2)
6. $A = 3.14(9)^2 = 254.34 \text{ ft}^2$
   $254.34 > 200$; it will not be enough.
7. $C = 2(3.14)(9) = 56.52 \text{ ft}$
   $56.522 \div 2 = 28.26$
   29 pieces must be bought
8. $C = 2(3.14)(100) = 628 \text{ miles}$

Systematic Review 16D

1. $A = 3.14(1.4)^2 = 6.1544 \text{ in}^2$
   $C = 2(3.14)(1.4) = 8.792 \text{ in}$
2. $A = 3.14(10)^2 = 314 \text{ ft}^2$
   $C = 3.14(20) = 62.8 \text{ ft}$

3. .15 m
4. .8 kl
5. 3,500 cg
6. $1.65 \times 9.43 = 15.5595$
7. $.209 \times 8.07 = 1.68663$
8. $5.061 \times 3.94 = 19.94034$
9. $6\frac{1}{5} \div 1\frac{3}{4} = \frac{31}{5} \div \frac{7}{4} = \frac{31}{5} \times \frac{4}{7} =$
   $\frac{124}{35} = 3\frac{19}{35}$
10. $4\frac{3}{5} \div 4\frac{1}{2} = \frac{23}{5} \div \frac{9}{2} = \frac{23}{5} \times \frac{2}{9} =$
    $\frac{46}{45} = 1\frac{1}{45}$
11. $\frac{1}{2} \div \frac{1}{8} = \frac{1}{2} \times \frac{\cancel{8}^4}{1} = 4$
12. done
13. $10'+10'+10'+10' = 40'$
14. $7+15+7+15 = 44 \text{ m}$
15. $12+13+12+13 = 50 \text{ m}$
16. $13+12+13 = 38 \text{ m}$
17. $38 \times 3 = 114 \text{ ft}$ (approximately)
18. $\$18,550.00 \times .02 = \$371.00$

Systematic Review 16E

1. $A = 3.14(3.2)^2 = 32.1536 \text{ cm}^2$
   $C = 2(3.14)(3.2) = 20.096 \text{ cm}$
2. $A = 3.14(2)^2 = 12.56 \text{ m}^2$
   $C = 3.14(4) = 12.56 \text{ m}$
3. .01 km
4. 5 cl
5. 20,000 dg
6. $.2 \times .32 = .064$
7. $1.45 \times .04 = .058$
8. $.005 \times .02 = .0001$
9. $3\frac{1}{2} \div 2\frac{1}{6} = \frac{7}{2} \div \frac{13}{6} =$
   $\frac{7}{2} \times \frac{\cancel{6}^3}{13} = \frac{21}{13} = 1\frac{8}{13}$

SYSTEMATIC REVIEW 16E - LESSON PRACTICE 17A

10. $5\frac{1}{4} \div 3\frac{1}{4} = \frac{21}{4} \div \frac{13}{4} =$
    $\frac{21}{\cancel{4}} \times \frac{\cancel{4}}{13} = \frac{21}{13} = 1\frac{8}{13}$

11. $\frac{7}{8} \div \frac{1}{16} = \frac{7}{\cancel{8}} \times \frac{\cancel{16}^2}{1} = 14$

12. $4 + 10 + 4 + 10 = 28"$

13. $45 + 45 + 45 + 45 = 180'$

14. $2.3 + 1.2 + 2.3 + 1.2 = 7$ m

15. $\$2,000.00 \times .23 = \$460.00$

16. $3.14(7)^2 = 153.86$ in^2
    (round to 154 in^2)

17. $50 + 75 + 50 + 75 = 250$ m

18. $\$12.35 \times 250 = \$3,087.50$

19. 4 kl

20. $3\frac{1}{2} \div \frac{1}{4} = \frac{7}{2} \div \frac{1}{4} =$
    $\frac{7}{\cancel{2}} \times \frac{\cancel{4}^2}{1} = 14$ containers

Systematic Review 16F

1. $A = 3.14(1)^2 = 3.14$ m^2
   $C = 2(3.14)(1) = 6.28$ m

2. $A = 3.14(.5)^2 = .785$ km^2
   $C = 3.14(1) = 3.14$ km

3. 1.7 cm

4. .32 L

5. 5,000 g

6. $.06 \times .05 = .003$

7. $45.1 \times .8 = 36.08$

8. $116 \times .19 = 22.04$

9. $7\frac{1}{8} \div 3\frac{3}{4} = \frac{57}{8} \div \frac{15}{4} =$
   $\frac{\cancel{57}^{19}}{\cancel{8}_2} \times \frac{\cancel{4}}{\cancel{15}_5} = \frac{19}{10} = 1\frac{9}{10}$

10. $2\frac{5}{6} \div 1\frac{1}{6} = \frac{17}{6} \div \frac{7}{6} =$
    $\frac{17}{\cancel{6}} \times \frac{\cancel{6}}{7} = \frac{17}{7} = 2\frac{3}{7}$

11. $\frac{5}{8} \div \frac{1}{4} = \frac{5}{\cancel{8}_2} \times \frac{\cancel{4}}{1} = \frac{5}{2} = 2\frac{1}{2}$

12. $6 + 2 + 6 + 2 = 16$ m

13. $1.2 + 1.2 + 1.2 + 1.2 = 4.8'$

14. $17 + 10 + 17 + 10 = 54"$

15. $\$10.95 \times 1.20 = \$13.14$

16. $3.14(15)^2 = 706.5$ in^2

17. $\$20.50 + \$15.98 + \$11.24 = \$47.72$
    $\$47.72 \times 1.06 = \$50.58$

18. $2\frac{1}{2} \times 1\frac{1}{2} = \frac{5}{2} \times \frac{3}{2} = \frac{15}{4} = 3\frac{3}{4}$ cups

19. $8\frac{1}{4} \div \frac{3}{4} = \frac{\cancel{33}^{11}}{\cancel{4}} \times \frac{\cancel{4}}{\cancel{3}} = 11$ gifts

20. 100 m = 100,000 mm = .1 km

Lesson Practice 17A

1. done

2. done

3. $\begin{array}{r} .587 \\ 6\overline{)3.522} \\ \underline{30} \\ 52 \\ \underline{48} \\ 42 \\ \underline{42} \end{array}$

4. $6 \times .587 = 3.522$

5. $\begin{array}{r} 8.8 \\ 9\overline{)79.2} \\ \underline{72} \\ 72 \\ \underline{72} \end{array}$

6. $9 \times 8.8 = 79.2$

7. $\begin{array}{r} .8 \\ 5\overline{)4.0} \\ \underline{40} \end{array}$

8. $5 \times .8 = 4$

9. $\begin{array}{r} .004 \\ 2\overline{).008} \\ \underline{8} \end{array}$

10. $2 \times .004 = .008$

LESSON PRACTICE 17A - LESSON PRACTICE 17C

11. .0125
    4) .0500
       4
       10
        8
       20

12. 4 × .0125 = .05
13. 3.6 ÷ 4 = .9 pies
14. 5 ÷ 8 = .625 pints
15. $0.75 ÷ 5 = $0.15

Lesson Practice 17B

1. 1.5
   4) 6.0
      4
      20
      20

2. 4 × 1.5 = 6
3. .09
   3) .27
       .27

4. 3 × .09 = .27
5. .006
   7) .042
       42

6. 7 × .006 = .042
7. .05
   8) .40
       .40

8. 8 × .05 = .4
9. .203
   6) 1.218
       12
       018
       18

10. 6 × .203 = 1.218

11. 42.8
    2) 85.6
       8
       5
       4
       16
       16

12. 2 × 42.8 = 85.6
13. 24.8 ÷ 2 = 12.4 pieces
    12 pieces with .4 of a
    piece left over
14. 2.4 ÷ 3 = .8 bushels
15. 34.4 × 4 = 137.6 qt jars

Lesson Practice 17C

1. .08
   5) .40
       40

2. 5 × .08 = .4
3. .068
   6) .408
       36
       48
       48

4. 6 × .068 = .408
5. .008
   8) .064
       64

6. 8 × .008 = .064
7. .3
   4) .12
       12

8. 4 × .3 = .12
9. .5
   6) 3.0
       30

10. 6 × .5 = 3

LESSON PRACTICE 17C - SYSTEMATIC REVIEW 17E

11.
$$\begin{array}{r} 3.12 \\ 9\overline{)28.08} \\ \underline{27} \\ 10 \\ \underline{9} \\ 18 \\ \underline{18} \end{array}$$

12. $9 \times 3.12 = 28.08$
13. $345.5 \div 2$ days = 172.75 miles per day
14. $125.1 \div 9$ days = 13.9 bushels
15. $\$24.90 \div 3 = \$8.30$ per child

Systematic Review 17D

1.
$$\begin{array}{r} .02 \\ 7\overline{).14} \\ \underline{.14} \end{array}$$

2. $7 \times .02 = .14$

3.
$$\begin{array}{r} 1.2 \\ 3\overline{)3.6} \\ \underline{3} \\ 6 \\ \underline{6} \end{array}$$

4. $3 \times 1.2 = 3.6$

5.
$$\begin{array}{r} .061 \\ 9\overline{).549} \\ \underline{54} \\ 9 \\ \underline{9} \end{array}$$

6. $9 \times .061 = .549$

7.
$$\begin{array}{r} .625 \\ 8\overline{)5.000} \\ \underline{4\ 8} \\ 20 \\ \underline{16} \\ 40 \\ \underline{40} \end{array}$$

8. $8 \times .625 = 5$
9. $3.14(2.5 \text{ mi})^2 = 19.625 \text{ mi}^2$
10. $3.14(5 \text{ mi}) = 15.7 \text{ mi}$

11. .015 hg
12. .005 kl
13. 160 cm
14. done
15. $13 + 3 + 13 + 3 = 32'$
16. $7.2 + 4.1 + 7.2 + 4.1 = 22.6$ m
17. $\$32.50 \div 2 = \$16.25$
18. $2\frac{3}{4} - \frac{5}{8} = 2\frac{6}{8} - \frac{5}{8} = 2\frac{1}{8}$ pizza

Systematic Review 17E

1.
$$\begin{array}{r} .008 \\ 4\overline{).032} \\ \underline{32} \end{array}$$

2. $4 \times .008 = .032$

3.
$$\begin{array}{r} .05 \\ 6\overline{).30} \\ \underline{30} \end{array}$$

4. $6 \times .05 = .3$

5.
$$\begin{array}{r} 11.1 \\ 5\overline{)55.5} \\ \underline{5} \\ 5 \\ \underline{5} \\ 5 \\ \underline{5} \end{array}$$

6. $5 \times 11.1 = 55.5$

7.
$$\begin{array}{r} .515 \\ 2\overline{)1.030} \\ \underline{1\ 0} \\ 3 \\ \underline{2} \\ 10 \\ \underline{10} \end{array}$$

8. $2 \times .515 = 1.030$
9. $3.14(8 \text{ km})^2 = 200.96 \text{ km}^2$
10. $2(3.14)(8 \text{ km}) = 50.24 \text{ km}$
11. 1 g
12. .003 km
13. 2.5 dl
14. $4 + 3.8 + 4 + 3.8 = 15.6$ m

SYSTEMATIC REVIEW 17E - LESSON PRACTICE 18A

15. $25 + 9.1 + 25 + 9.1 = 68.2"$
16. $.12 + .05 + .12 + .05 = .34$ km
17. $\$15.00 \times .9 = \$13.50$ sale price
18. $\$13.50 \times 1.04 = \$14.04$ total cost
    $\$15.00 - \$14.04 = \$0.96$ left over
19. $5.7 \div 3 = 1.9$ km
20. $10 - 2\frac{5}{8} = 9\frac{8}{8} - 2\frac{5}{8} = 7\frac{3}{8}$ lb

17. $5,000$ m $= 5$ km $= 500,000$ cm
18. $\$5.76 = \$0.89 = \$4.87$
19. $\$38.40 \times 1.15 = \$44.16$ total
    $\$44.16 \div 4 = \$11.04$ each
20. $5\frac{1}{2} - 2\frac{2}{3} = 5\frac{3}{6} - 2\frac{4}{6} =$
    $4\frac{9}{6} - 2\frac{4}{6} = 2\frac{5}{6}$ rows

Systematic Review 17F

1. $.001$
    $1\overline{)\,.001}$
    $\underline{1}$
2. $1 \times .001 = .001$
3. $.32$
    $7\overline{)2.24}$
    $\underline{2\ 1}$
    14
    $\underline{14}$
4. $7 \times .32 = 2.24$
5. $.05$
    $3\overline{)\,.15}$
    $\underline{15}$
6. $3 \times .05 = .15$
7. $.125$
    $8\overline{)1.000}$
    $\underline{8}$
    20
    $\underline{16}$
    40
    $\underline{40}$
8. $8 \times .125 = 1$
9. $3.14(2 \text{ mi})^2 = 12.56 \text{ mi}^2$
10. $2(3.14)(2 \text{ mi}) = 12.56$ mi
11. $.95$ hl
12. 2 cm
13. 40 cg
14. $80 + 70 + 80 + 70 = 300"$
15. $3.1 + 1.3 + 3.1 + 1.3 = 8.8$ m
16. $5\frac{1}{2} + 3\frac{1}{2} + 5\frac{1}{2} + 3\frac{1}{2} = 18"$

Lesson Practice 18A

1. done
2. done
3. 300
    $6\overline{)1800}$
    $\underline{1800}$
4. $.06 \times 300 = 18$
5. $1,000$
    $5\overline{)5,000}$
    $\underline{5,000}$
6. $.5 \times 1,000 = 500$
7. $9,000$
    $7\overline{)63,000}$
    $\underline{63,000}$
8. $.007 \times 9,000 = 63$
9. 120
    $9\overline{)1,080}$
    $\underline{9}$
    180
    $\underline{180}$
10. $.9 \times 120 = 108$
11. $1,600$
    $25\overline{)40,000}$
    $\underline{25}$
    $15,000$
    $\underline{15,000}$
12. $.25 \times 1,600 = 400$

LESSON PRACTICE 18A - LESSON PRACTICE 18C

13.
$$\begin{array}{r} 80 \\ 5\overline{\smash{)}400} \\ \underline{400} \end{array}$$
    $40 \div .5 = 80$ people

14.
$$\begin{array}{r} 80 \\ 3\overline{\smash{)}240} \\ \underline{240} \end{array}$$
    $24 \div .3 = 80$ days

15.
$$\begin{array}{r} 30 \\ 8\overline{\smash{)}240} \\ \underline{240} \end{array}$$
    $24 \div .8 = 30$ days

Lesson Practice 18B

1.
$$\begin{array}{r} 6,320 \\ 1\overline{\smash{)}6,320} \\ \underline{6,320} \end{array}$$

2. $.1 \times 6,320 = 632$

3.
$$\begin{array}{r} 2,200 \\ 15\overline{\smash{)}33,000} \\ \underline{30} \\ 3,000 \\ \underline{3,000} \end{array}$$

4. $.15 \times 2,200 = 330$

5.
$$\begin{array}{r} 230 \\ 6\overline{\smash{)}1,380} \\ \underline{1\,2} \\ 180 \\ \underline{180} \end{array}$$

6. $.6 \times 230 = 138$

7.
$$\begin{array}{r} 2,000 \\ 33\overline{\smash{)}66,000} \\ \underline{66,000} \end{array}$$

8. $.033 \times 2,000 = 66$

9.
$$\begin{array}{r} 530 \\ 4\overline{\smash{)}2,120} \\ \underline{2,000} \\ 120 \\ \underline{120} \end{array}$$
    (Filling in extra zeros to mark place value is optional when working a division problem.)

10. $.4 \times 530 = 212$

11.
$$\begin{array}{r} 500 \\ 75\overline{\smash{)}37,500} \\ \underline{37,500} \end{array}$$

12. $.75 \times 500 = 375$

13. $\$200 \div \$0.50 = 400$ gifts

14. $36 \div .9 = 40$ pillows

15. $3 \div .01 = 300$ days

Lesson Practice 18C

1.
$$\begin{array}{r} 2,290 \\ 2\overline{\smash{)}4,580} \\ \underline{4,000} \\ 580 \\ \underline{400} \\ 180 \\ \underline{180} \end{array}$$

2. $.2 \times 2,290 = 458$

3.
$$\begin{array}{r} 200 \\ 8\overline{\smash{)}1,600} \\ \underline{1,600} \end{array}$$

4. $.08 \times 200 = 16$

5.
$$\begin{array}{r} 160 \\ 7\overline{\smash{)}1,120} \\ \underline{700} \\ 420 \\ \underline{420} \end{array}$$

6. $.7 \times 160 = 112$

LESSON PRACTICE 18C - SYSTEMATIC REVIEW 18E

7. 38,000
    9 ⟌ 342,000
     270,000
     72,000
     72,000

8. .009 × 38,000 = 342

9. 1,190
    3 ⟌ 3,570
     3,000
     570
     300
     270
     270

10. .3 × 1,190 = 357

11. 1,000
    45 ⟌ 45,000
     45,000

12. .45 × 1,000 = 450
13. 5 ÷ .05 = 100 batches
14. \$35.00 ÷ \$0.25 = 140 quarters
15. 224 ÷ 4 = 560 customers

Systematic Review 18D

1. 60
    6 ⟌ 360
     360

2. .6 × 60 = 36

3. 2,000
    9 ⟌ 18,000
     18,000

4. .09 × 2,000 = 180

5. 3.54
    2 ⟌ 7.08
     6 00
     10
     10
     8
     8

6. 2 × 3.54 = 7.08

7. .002
    15 ⟌ .030
     30

8. 15 × .002 = .03
9. $3.14(5.5 \text{ ft})^2 = 94.985 \text{ ft}^2$
10. 3.14(11 ft) = 34.54 ft
11. 8 × 8 = 64
12. 10 × 10 = 100
13. 2 × 2 = 4
14. done
15. 3 + 4 + 5 = 12'
16. 5.4 + 3.9 + 7.8 = 17.1'
17. 14 + 14 + 14 + 14 = 56"
18. 5 × 1.50 = 7.5 tons

Systematic Review 18E

1. 1,000
    1 ⟌ 1,000
     1,000

2. .1 × 1,000 = 100

3. 250
    6 ⟌ 1,500
     1 2
     300
     300

4. .06 × 250 = 15

5. 1.9
    8 ⟌ 15.2
     8 0
     7 2
     7 2

6. 8 × 1.9 = 15.2

7. .0002
    30 ⟌ .0060
     60

8. 30 × .0002 = .006
9. $3.14(9 \text{ m})^2 = 254.34 \text{ m}^2$
10. 2(3.14)(9 m) = 56.52 m
11. 12 × 12 = 144
12. 1 × 1 = 1

13. $5 \times 5 = 25$
14. $8 + 8 + 10 = 26'$
15. $5 + 12 + 13 = 30"$
16. $1.7 + 1.1 + 1.9 = 4.7$ m
17. $20 + 30 + 20 + 30 = 100$ cm $= 1$ m
18. $100 \times 2.1 = 210$ nature books
19. $\frac{1}{2} + 1\frac{1}{4} = \frac{2}{4} + 1\frac{1}{4} = 1\frac{3}{4}$ cups
20. $\$8.00 \div \$0.75 = 10.67 = 10$ cones

Systematic Review 18F

1. $\overline{123,000}$
    $3\,\overline{)369,000}$
    $\underline{300,000}$
    $69,000$
    $\underline{60,000}$
    $9,000$
    $\underline{9,000}$
2. $.003 \times 123,000 = 369$
3. $\overline{100}$
    $49\,\overline{)4,900}$
    $\underline{4,900}$
4. $.49 \times 100 = 49$
5. $\overline{4.5}$
    $5\,\overline{)22.5}$
    $\underline{200}$
    25
    $\underline{25}$
6. $5 \times 4.5 = 22.5$
7. $\overline{.0002}$
    $8\,\overline{).0016}$
    $\underline{16}$
8. $8 \times .0002 = .0016$
9. $3.14(1.2 \text{ in})^2 = 4.5216 \text{ in}^2$
10. $2(3.14)(1.2 \text{ in}) = 7.536$ in
11. $25 \times 25 = 625$
12. $7 \times 7 = 49$
13. $100 \times 100 = 10,000$
14. $25 + 25 + 32 = 82'$
15. $6 + 8 + 10 = 24"$
16. $.89 + .7 + 1.08 = 2.67$ m
17. $25 \div .5 = 50$ packages
18. $\$37.50 \times 1.50 = \$56.25$
    $\$56.25 \div 50 = \$1.125 /$ package
    rounds to $\$1.13$ per package
19. $\$42.50 \times 1.34 = \$56.95$
20. $4,520 = 4.52$ kg

Lesson Practice 19A

1. done
2. done
3. $.06X = 24$
    $\dfrac{.\cancel{06}X}{.\cancel{06}} = \dfrac{24}{.06}$
    $X = 24 \div .06 = 400$
4. $.06(400) = 24$
    $24 = 24$
5. $.7A = 490$
    $\dfrac{.\cancel{7}A}{.\cancel{7}} = \dfrac{490}{.7}$
    $A = 490 \div .7 = 700$
6. $.7(700) = 490$
    $490 = 490$
7. $.002R = 6$
    $\dfrac{.\cancel{002}R}{.\cancel{002}} = \dfrac{6}{.002}$
    $R = 6 \div .002 = 3,000$
8. $.002(3,000) = 6$
    $6 = 6$
9. $.2M = 84$
    $\dfrac{.\cancel{2}M}{.\cancel{2}} = \dfrac{84}{.2}$
    $M = 84 \div .2 = \$420.00$
10. $.15T = 3$
    $\dfrac{.\cancel{15}T}{.\cancel{15}} = \dfrac{3}{.15}$
    $T = 3 \div .15 = 20'$

Lesson Practice 19B

1. $.5Y = 8$

    $\dfrac{.5Y}{.5} = \dfrac{8}{.5}$

    $Y = 8 \div .5 = 16$

2. $.5(16) = 8$

    $8 = 8$

3. $.03X = 21$

    $\dfrac{.03X}{.03} = \dfrac{21}{.03}$

    $X = 21 \div .03 = 700$

4. $.03(700) = 21$

    $21 = 21$

5. $.1A = 30$

    $\dfrac{.1A}{.1} = \dfrac{30}{.1}$

    $A = 30 \div .1 = 300$

6. $.1(300) = 30$

    $30 = 30$

7. $.008R = 16$

    $\dfrac{.008R}{.008} = \dfrac{16}{.008}$

    $R = 16 \div .008 = 2,000$

8. $.008(2,000) = 16$

    $16 = 16$

9. $.7P = 140$

    $\dfrac{.7P}{.7} = \dfrac{140}{.7}$

    $P = 140 \div .7 = \$200.00$

10. $.8H = 60$

    $\dfrac{.8H}{.8} = \dfrac{60}{.8}$

    $H = 60 \div .8 = 75"$

Lesson Practice 19C

1. $.6Y = 6$

    $\dfrac{.6Y}{.6} = \dfrac{6}{.6}$

    $Y = 6 \div .6 = 10$

2. $.6(10) = 6$

    $6 = 6$

3. $.04X = 32$

    $\dfrac{.04X}{.04} = \dfrac{32}{.04}$

    $X = 32 \div .04 = 800$

4. $.04(800) = 32$

    $32 = 32$

5. $.3A = 93$

    $\dfrac{.3A}{.3} = \dfrac{93}{.3}$

    $A = 93 \div .3 = 310$

6. $.3(310) = 93$

    $93 = 93$

7. $.005R = 5$

    $\dfrac{.005R}{.005} = \dfrac{5}{.005}$

    $R = 5 \div .005 = 1,000$

8. $.005(1,000) = 5$

    $5 = 5$

9. $.25G = 4$

    $\dfrac{.25G}{.25} = \dfrac{4}{.25}$

    $G = 4 \div .25 = 16$ bushels

10. $.05R = 10$

    $\dfrac{.05R}{.05} = \dfrac{10}{.05}$

    $R = 10 \div .05 = \$200.00$

Systematic Review 19D

1. $.22Q = 88$

    $\dfrac{.22Q}{.22} = \dfrac{88}{.22}$

    $Y = 88 \div .22 = 400$

2. $.22(400) = 88$

    $88 = 88$

3. $.007X = 35$

    $\dfrac{.007X}{.007} = \dfrac{35}{.007}$

    $X = 35 \div .007 = 5,000$

SYSTEMATIC REVIEW19D - SYSTEMATIC REVIEW 19F

4. $.007(5,000) = 35$
   $35 = 35$
5. $4 \div 20 = .2$
6. $20 \times .2 = 4$
7. $3.6 \div 18 = .2$
8. $18 \times .2 = 3.6$
9. done
10. $10 \times 10 = 100$ ft^2
11. $15 \times 7 = 105$ m^2
12. $9 \times 12 = 108$ ft^2
13. $9 + 12 + 9 + 12 = 42'$
14. $.35M = 105$
    $\dfrac{.35M}{.35} = \dfrac{105}{.35}$
    $M = 105 \div .35 = \$300.00$
15. $2(3.14)(5 \text{ mi}) = 31.4$ mi
16. $3.14(5 \text{ mi})^2 = 78.5$ mi^2
17. 5 km $= 5,000$ m
    $5,000$ m > 500 m; She drove further.
18. $\$30.75 \div 5 = \$6.15$

10. $45 \times 45 = 2,025$ ft^2
11. $2.3 \times 1.2 = 2.76$ m^2
12. $4 \times 11 = 44$ yd^2
13. $3 + 3 + 3 + 3 = 12$ ft
    $12 \times 1 = 12$ flowers
14. $.40P = 80$
    $\dfrac{.40P}{.40} = \dfrac{80}{.40}$
    $P = 80 \div .40 = \$200.00$
15. $10 \times 12 = 120$ in^2
16. $3.14(6 \text{ in})^2 = 113.04$ in^2
17. 120 in$^2 > 113.04$ in^2
    The rectangular one would be better.
18. 2 m $= 200$ cm
    They are the same height.
19. $\$57.04 \div 8 = \$7.13$
20. $4.5 + 5 + 3.75 = 13.25"$

Systematic Review 19E

1. $.9D = 27$
   $\dfrac{.9D}{.9} = \dfrac{27}{.9}$
   $D = 27 \div .9 = 30$
2. $.9(30) = 27$
   $27 = 27$
3. $.08F = 5$
   $\dfrac{.08F}{.08} = \dfrac{5}{.08}$
   $F = 5 \div .08 = 62.5$
4. $.08(62.5) = 5$
   $5 = 5$
5. $5,500 \div 11 = 500$
6. $.11 \times 500 = 55$
7. $.16 \div 40 = .004$
8. $40 \times .004 = .16$
9. $10 \times 4 = 40$ in^2

Systematic Review 19F

1. $.32G = 64$
   $\dfrac{.32G}{.32} = \dfrac{64}{.32}$
   $G = 64 \div .32 = 200$
2. $.32(200) = 64$
   $64 = 64$
3. $.005Y = 4$
   $\dfrac{.005Y}{.005} = \dfrac{4}{.005}$
   $Y = 4 \div .005 = 800$
4. $.005(800) = 4$
   $4 = 4$
5. $10,000 \div 1 = 10,000$
6. $.001 \times 10,000 = 10$
7. $1.26 \div 21 = .06$
8. $.06 \times 21 = 1.26$
9. $6 \times 2 = 12$ m^2
10. $1.2 \times 1.2 = 1.44$ ft^2
11. $17 \times 10 = 170$ in^2

SYSTEMATIC REVIEW 19F - LESSON PRACTICE 20B

12. $5 \times 6 = 30$ ft^2
13. $3.14(3 \text{ ft})^2 = 28.26$ ft^2
14. $30 - 28.26 = 1.74$ ft^2
15. $1 + 1 + 1 + 1 = 4$ mi
16. $.32T = 200$

    $\dfrac{.32T}{.32} = \dfrac{200}{.32}$

    $T = 200 \div .32 = 625$ hr
17. $625 - 200 = 425$ hr
18. 652 g $= .652$ kg
19. $\$121.92 \div 12 = \$10.16$
20. $6.25 + 7.1 + 6.25 + 7.1 = 26.7"$

Lesson Practice 20A

1. done
2. done
3. done
4. done
5. 9
   $9\overline{)81}$
   $\underline{81}$
6. $.9 \times 9 = 8.1$
7. 40
   $6\overline{)240}$
   $\underline{240}$
8. $.006 \times 40 = .24$
9. 59
   $8\overline{)472}$
   $\underline{472}$
10. $.8 \times 59 = 47.2$
11. 2.72
    $15\overline{)40.80}$
    $\underline{30\ 00}$
    $10\ 80$
    $\underline{10\ 50}$
    30
    $\underline{30}$
12. $1.5 \times 2.72 = 4.080$
13. $1.8 \div .3 = 6$ guests

14. $2.6 \div 1.3 = 2$ days
15. $6.9 \div 2.3 = 3$ people

Lesson Practice 20B

1. $.09$
   $5\overline{).45}$
   $\underline{.45}$
2. $.5 \times .09 = .045$
3. 5
   $82\overline{)410}$
   $\underline{410}$
4. $.82 \times 5 = 4.1$
5. 1.1
   $2\overline{)2.2}$
   $\underline{2\ 0}$
   2
   $\underline{2}$
6. $.2 \times 1.1 = .22$
7. 71
   $9\overline{)639}$
   $\underline{630}$
   9
   $\underline{9}$
8. $.009 \times 71 = .639$
9. 61.2
   $7\overline{)428.4}$
   $\underline{420\ 0}$
   $8\ 4$
   $\underline{7\ 0}$
   $1\ 4$
   $\underline{1\ 4}$
10. $.7 \times 61.2 = 42.84$
11. 15
    $21\overline{)315}$
    $\underline{210}$
    105
    $\underline{105}$
12. $2.1 \times 15 = 31.5$

LESSON PRACTICE 20B - SYSTEMATIC REVIEW 20D

13. $.856 \div .4 = 2.14$ sacks
    He has 3 sacks.
14. $5.5 \div .25 = 22$ favors
15. $\$45.50 \div \$0.35 = 130$ items

13. $25.5 \div .5 = 51$ bags
14. $4.6 \div .2 = 23$ rests
15. $\$16.80 \div \$2.10 = 8$ items

Lesson Practice 20C

1. $.06$
    $8\overline{).48}$
    $\underline{.48}$

2. $.8 \times .06 = .048$

3. 40.6
    $22\overline{)893.2}$
    $\underline{8800}$
    $13\,2$
    $\underline{13\,2}$

4. $.22 \times 40.6 = 8.932$

5. 13
    $5\overline{)65}$
    $\underline{50}$
    15
    $\underline{15}$

6. $.5 \times 13 = 6.5$

7. 580
    $7\overline{)4060}$
    $\underline{3500}$
    560
    $\underline{560}$

8. $.007 \times 580 = 4.06$

9. 91.2
    $4\overline{)364.8}$
    $\underline{360\,0}$
    $4\,8$
    $\underline{4\,8}$

10. $.4 \times 91.2 = 36.48$

11. 16
    $32\overline{)512}$
    $\underline{320}$
    192
    $\underline{192}$

12. $3.2 \times 16 = 51.2$

Systematic Review 20D

1. $.5$
    $3\overline{)1.5}$
    $\underline{1\,5}$

2. $.3 \times .5 = .15$

3. 1.6
    $25\overline{)40.0}$
    $\underline{25\,0}$
    $15\,0$
    $\underline{15\,0}$

4. $.025 \times 1.6 = .04$

5. $.002$
    $60\overline{).120}$
    $\underline{120}$

6. $60 \times .002 = .12$

7. 430
    $1\overline{)430}$
    $\underline{400}$
    30
    $\underline{30}$

8. $.1 \times 430 = 43$

9. $.8G = 40$
    $G = 40 \div .8 = 50$

10. $.15Y = .3$
    $Y = .3 \div .15 = 2$

11. $8\frac{1}{2} + 2\frac{7}{8} = 8\frac{4}{8} + 2\frac{7}{8} = 10\frac{11}{8} = 11\frac{3}{8}$

12. $16 - 1\frac{2}{3} = 15\frac{3}{3} - 1\frac{2}{3} = 14\frac{1}{3}$

13. $3\frac{1}{8} \times 1\frac{1}{3} = \frac{25}{\cancel{8}_2} \times \frac{\cancel{4}}{3} = \frac{25}{6} = 4\frac{1}{6}$

14. done

15. $13 \times 28 = 364 \text{ in}^2$

16. $2 \times 1.2 = 2.4 \text{ m}^2$

17. $220+150+220+150=740$ m
    $220\times100=22{,}000$ m^2
18. $.25M=37.50$
    $M=37.50\div.25=\$150.00$

Systematic Review 20E

1.
$$\begin{array}{r}228\\2\overline{)456}\\\underline{400}\\56\\\underline{40}\\16\\\underline{16}\end{array}$$

2. $.02\times 228=4.56$
3.
$$\begin{array}{r}910\\9\overline{)8{,}190}\\\underline{8{,}100}\\90\\\underline{90}\end{array}$$

4. $.009\times 910=8.19$
5.
$$\begin{array}{r}5.01\\12\overline{)60.12}\\\underline{60\ 00}\\12\\\underline{12}\end{array}$$

6. $12\times 5.01=60.12$
7.
$$\begin{array}{r}20\\45\overline{)900}\\\underline{900}\end{array}$$

8. $4.5\times 20=90$
9. $6.2X=7.44$
   $X=7.44\div 6.2=1.2$
10. $.4B=.88$
    $B=.88\div .4=2.2$
11. $6\frac{3}{8}\div 3\frac{1}{4}=\frac{51}{8}\div\frac{13}{4}=$
    $\frac{51}{\cancel{8}_2}\times\frac{\cancel{4}}{13}=\frac{51}{26}=1\frac{25}{26}$
12. $65-21\frac{5}{6}=64\frac{6}{6}-21\frac{5}{6}=43\frac{1}{6}$

13. $9\frac{3}{10}\times 5\frac{2}{3}=\frac{\cancel{93}^{31}}{10}\times\frac{17}{\cancel{3}}=$
    $\frac{527}{10}=52\frac{7}{10}$
14. $3.8\times 3=11.4$ m^2
15. $20\times 7=140$ mm^2
16. $25\times 8.5=212.5$ in^2
17. $25\times 33=825$ yd^2 total
    $825\times\frac{1}{3}=275$ yd^2 planted
18. $275\div .5=550$ plants
19. $550\times \$0.85=\$467.50$
20. $1.05A=67.41$
    $A=67.41\div 1.05=\$64.20$

Systematic Review 20F

1.
$$\begin{array}{r}78\\4\overline{)312}\\\underline{280}\\32\\\underline{32}\end{array}$$

2. $.4\times 78=31.2$
3.
$$\begin{array}{r}15\\28\overline{)420}\\\underline{280}\\140\\\underline{140}\end{array}$$

4. $.28\times 15=4.2$
5.
$$\begin{array}{r}.056\\35\overline{)1.960}\\\underline{1\ 750}\\210\\\underline{210}\end{array}$$

6. $35\times .056=1.96$
7.
$$\begin{array}{r}20\\88\overline{)1760}\\\underline{1760}\end{array}$$

8. $8.8\times 20=176$
9. $.11Q=.55$
   $Q=.55\div .11=5$

SYSTEMATIC REVIEW 20F - LESSON PRACTICE 21A

10. $1.3W = .39$
    $W = .39 \div 1.3 = .3$

11. $\dfrac{6}{5} \div \dfrac{2}{5} = \dfrac{6 \div 2}{1} = 3$

12. $18 + 6\dfrac{7}{8} = 24\dfrac{7}{8}$

13. $\dfrac{\cancel{11}}{\cancel{12}_{3}} \times \dfrac{\cancel{4}}{5} = \dfrac{11}{15}$

14. $80 \times 55 = 4,400 \text{ in}^2$

15. $3.1 \times .9 = 2.79 \text{ m}^2$

16. $\dfrac{11}{2} \times \dfrac{21}{8} = \dfrac{231}{16} = 14\dfrac{7}{16}$

17. $3.14(6^2) = 113.04 \text{ ft}^2$

18. $5 \times 1,000 = 5,000 \text{ m}$

19. $1.15M = 18.86$
    $M = 18.86 \div 1.15 = \$16.40$

20. $9 + 12 + 15 = 36 \text{ mi}$

5. $1.973 \approx 1.97$
    ```
      3 ) 5.920
          3 000
          2 920
          2 700
            220
            210
             10
    ```
    ($\approx$ means "approximately equal to")

6. $17.555 \approx 17.56$
    ```
      9 ) 158.000
          90 000
          68 000
          63 000
           5 000
           4 500
             500
             450
              50
              45
               5
    ```

Lesson Practice 21A

1. done

2.  ```
          .712
      5 ) 3.560
          3 500
             60
             50
             10
             10
    ```

3.  ```
          11.25
      8 ) 90.00
          80 00
          10 00
           8 00
           2 00
           1 60
             40
             40
    ```

4. done

7. done

8. 7.33 or $7.\overline{3}$
    ```
      3 ) 22.00
          21 00
           1 00
             90
             10
              9
    ```

9. 9.166 or $9.1\overline{6}$
    ```
      6 ) 55.000
          54 000
           1 000
             600
             400
             360
              40
              36
    ```

10. done

11. $.59\frac{6}{8} = .59\frac{3}{4}$

    ```
    8 ) 4.78
        4 00
          78
          72
           6
    ```

12. $4.72\frac{2}{9}$

    ```
    9 ) 42.50
        36 00
         6 50
         6 30
           20
           18
    ```

Lesson Practice 21B

1.  ```
      .2335
    2 ) .4670
        4000
         670
         600
          70
          60
          10
          10
    ```

2.  ```
       12.75
    4 ) 51.00
        40 00
        11 00
         8 00
         3 00
          280
           20
           20
    ```

3.  ```
        15.4
    5 ) 77.0
        5 00
        2 70
        2 50
          20
          20
    ```

4. $4.557 \approx 4.56$

    ```
    7 ) 31.900
        28 000
         3 900
         3 500
           400
           350
            50
            49
    ```

5. $1.093 \approx 1.09$

    ```
    9 ) 9.840
        9 000
          840
          810
           30
           27
    ```

6. $50.833 \approx 50.83$

    ```
    6 ) 305.000
        300 000
          5 000
          4 800
            200
            180
             20
             18
    ```

7. $.133$ or $.1\overline{3}$

    ```
    3 ) .400
         300
         100
          90
          10
           9
    ```
    ```

LESSON PRACTICE 21B - LESSON PRACTICE 21C

8. .01428571 or .01$\overline{42857}$
```
   7 ).10000000
       7000000
       3000000
       2800000
        200000
        140000
         60000
         56000
          4000
          3500
           500
           490
            10
             7
```

9. 9.166 or 9.1$\overline{6}$
```
   6 )55.000
      54 000
       1 000
         600
         400
         360
          40
          36
```

10. 14.07$\frac{1}{7}$
```
    7 )98.50
       70 00
       28 50
       28 00
          50
          49
           1
```

11. 244.66$\frac{2}{3}$
```
    3 )734.00
       600 00
       134 00
       120 00
        14 00
        12 00
         2 00
         1 80
```

12. 13.87$\frac{4}{8}$ = 13.87$\frac{1}{2}$
```
    8 )111.00
       80 00
       31 00
       24 00
        7 00
        6 40
          60
          56
           4
```

Lesson Practice 21C

1. 7.1375
```
   8 )57.1000
      56 0000
       1 1000
         8000
         3000
         2400
          600
          560
           40
           40
```

2. 19.5
```
   2 )39.0
      200
      190
      180
       10
       10
```

3. 3.125
```
   4 )12.500
      12 000
         500
         400
         100
          80
          20
          20
```

LESSON PRACTICE 21C - SYSTEMATIC REVIEW 21D

4. $2.266 \approx 2.27$
    $3\overline{)6.800}$
    $\underline{6\ 000}$
    800
    $\underline{600}$
    200
    $\underline{180}$
    20
    $\underline{18}$

5. $.742 \approx .74$
    $7\overline{)5.200}$
    $\underline{4\ 900}$
    300
    $\underline{280}$
    20
    $\underline{14}$

6. $.535 \approx .54$
    $9\overline{)4.820}$
    $\underline{4\ 500}$
    320
    $\underline{270}$
    50
    $\underline{45}$

7. $.4077$ or $.40\overline{7}$
    $9\overline{)3.6700}$
    $\underline{3\ 6000}$
    700
    $\underline{630}$
    70
    $\underline{63}$

8. $.181$ or $.\overline{18}$
    $11\overline{)2.000}$
    $\underline{1\ 100}$
    900
    $\underline{880}$
    20
    $\underline{11}$

9. 15.0166 or $15.01\overline{6}$
    $6\overline{)90.1000}$
    $\underline{60\ 0000}$
    $30\ 1000$
    $\underline{30\ 0000}$
    1000
    $\underline{600}$
    400
    $\underline{360}$
    40
    $\underline{36}$

10. $.78\frac{2}{10} = .78\frac{1}{5}$
    $10\overline{)7.82}$
    $\underline{7\ 00}$
    82
    $\underline{80}$
    2

11. $.04\frac{5}{7}$
    $7\overline{).33}$
    $\underline{28}$
    5

12. $2.04\frac{2}{12} = 2.04\frac{1}{6}$
    $12\overline{)24.50}$
    $\underline{24\ 00}$
    50
    $\underline{48}$
    2

Systematic Review 21D

1. $.178 \approx .18$
    $7\overline{)1.250}$
    $\underline{700}$
    550
    $\underline{490}$
    60
    $\underline{56}$

SYSTEMATIC REVIEW 21D - SYSTEMATIC REVIEW 21D

2. $1.666 \approx 1.67$
    ```
    3 ) 5.000
        3 000
        2 000
        1 800
          200
          180
           20
           18
    ```

3. $5.483 \approx 5.48$
    ```
    6 ) 32.900
        30 000
         2 900
         2 400
           500
           480
            20
            18
    ```

4. $.9344$ or $.93\overline{4}$
    ```
    9 ) 8.4100
        8 1000
          3100
          2700
           400
           360
            40
            36
    ```

5. 6.33 or $6.\overline{3}$
    ```
    6 ) 38.00
        36 00
         2 00
         1 80
           20
           18
    ```

6. $.2857142$ or $.\overline{285714}$
    ```
    7 ) 2.0000000
        1 4000000
          6000000
          5600000
           400000
           350000
            50000
            49000
             1000
              700
              300
              280
               20
               14
    ```

7. $8X = .5$
    $X = .5 \div 8 = .06\frac{4}{8} = .06\frac{1}{4}$

8. $.3Y = 5.63$
    $Y = 5.63 \div .3 = 18.76\frac{2}{3}$

9. $10F = .46$
    $F = .46 \div 10 = .04\frac{6}{10} = .04\frac{3}{5}$

10. $.05 + 1.9 = 1.95$
11. $3.67 - .08 = 3.59$
12. $.006 + 4.2 = 4.206$
13. done
14. $(1/2)(9)(12) = 54$ in^2
15. $(1/2)(3.08)(.7) = 1.078$ m^2
16. $\$3.55 \times .20 = \$71.00$
17. $2,400 \div 1,000 = 2.4$ kg
    $25 - 2.4 = 22.6$ kg
18. $7H = 37.1$
    $H = 37.1 \div 7 = 5.3$ ft

Systematic Review 21E

1. $6.844 \approx 6.84$
    $9\,\overline{)61.600}$
    $\underline{54\,000}$
    $7\,600$
    $\underline{7\,200}$
    400
    $\underline{360}$
    40
    $\underline{36}$

2. $.018 \approx .02$
    $20\,\overline{).370}$
    $\underline{200}$
    170
    $\underline{160}$

3. $.133 \approx .13$
    $3\,\overline{).400}$
    $\underline{300}$
    100
    $\underline{90}$
    10
    $\underline{9}$

4. $.266$ or $.2\overline{6}$
    $6\,\overline{)1.600}$
    $\underline{1\,200}$
    400
    $\underline{360}$
    40
    $\underline{36}$

5. 23.33 or $23.\overline{3}$
    $3\,\overline{)70.00}$
    $\underline{60\,00}$
    $10\,00$
    $\underline{9\,00}$
    $1\,00$
    $\underline{90}$

6. $.10909$ or $.1\overline{09}$
    $11\,\overline{)1.20000}$
    $\underline{1\,10000}$
    10000
    $\underline{9900}$
    100
    $\underline{99}$

7. $12X = .28$
    $X = .28 \div 12 = .02\frac{4}{12} = .02\frac{1}{3}$

8. $.9Y = .5$
    $Y = .5 \div .9 = .55\frac{5}{9}$

9. $1.3F = 5.4$
    $F = 5.4 \div 1.3 = 4.15\frac{5}{13}$

10. $6.15 - .06 = 6.09$
11. $14.003 + .5 = 14.503$
12. $8.3 - .67 = 7.63$
13. $1/2(4)(6) = 12\,\text{ft}^2$
14. $1/2(12)(16) = 96\,\text{in}^2$
15. $1/2(1.02)(.5) = .255\,\text{m}^2$
16. $\$11.00 \times 1.06 = \$11.66$
17. $\$.96 \div 12 = \$.08$ per oz
18. $\$.56 \div 8 = \$.07$ per oz
    The 8-ounce can is a better buy.
19. $4{,}180 \div 1{,}000 = 4.18$ km
    $4.18 - 4 = .18$ km
20. $20H = 40.22$
    $H = 40.22 \div 20$
    $H = 2.011$ in (rounds to 2.01 in)

Systematic Review 21F

1. $8.062 \approx 8.06$
    $8\,\overline{)64.500}$
    $\underline{64\,000}$
    500
    $\underline{480}$
    20
    $\underline{16}$

2. $.304 \approx .30$
    $15\,\overline{)4.570}$
    $\underline{4\,500}$
    70
    $\underline{60}$

SYSTEMATIC REVIEW 21F - LESSON PRACTICE 22A

3. $2.128 \approx 2.13$
    $7\overline{)14.900}$
    $\underline{14\,000}$
    900
    $\underline{700}$
    200
    $\underline{140}$
    60
    $\underline{56}$

4. 2.242 or $2.2\overline{4}$
    $33\overline{)74.000}$
    $\underline{66\,000}$
    $8\,000$
    $\underline{6\,600}$
    $1\,400$
    $\underline{1\,320}$
    80
    $\underline{66}$

5. 50.909 or $50.9\overline{0}$
    $11\overline{)560.000}$
    $\underline{550\,000}$
    $10\,000$
    $\underline{9\,900}$
    100
    $\underline{99}$

6. 1.11 or $1.\overline{1}$
    $9\overline{)10.00}$
    $\underline{9\,00}$
    $1\,00$
    $\underline{90}$
    10
    $\underline{9}$

7. $40X = .6$
    $X = .6 \div 40 = .01\frac{1}{2}$

8. $.2Y = .077$
    $Y = .077 \div .2 = .38\frac{1}{2}$

9. $4.1F = 8.24$
    $F = 8.24 \div 4.1 = 2.00\frac{40}{41}$

10. $11 - .99 = 10.01$

11. $5.003 + 21 = 26.003$

12. $100 - .01 = 99.99$

13. $1/2(10)(12) = 60$ ft^2

14. $1/2(7)(24) = 84$ in^2

15. $1/2(1.3)(4.2) = 2.73$ m^2

16. $\$45.00 \times 1.08 = \$48.60$

17. $\$100.00 \times 3.15 = \$315.00$

18. $3.14(7)^2 = 153.86$ in^2
    $\$10.00 \div 153.86 = \$0.064994...$
    (rounds to $\$0.06$ per in^2)

19. $3.14(6)^2 = 113$ in^2
    $\$8.00 \div 113 = \$.07$ per in^2
    14 in is the better buy.

20. $4.25 \times 10 = 42.5$ mm
    $425 - 42.5 = 382.5$ mm

Lesson Practice 22A

1. done

2. done

3. $.09X + 4 = 4.9$
    $.09X = 4.9 - 4$
    $.09X = .9$
    $X = .9 \div .09$
    $X = 10$

4. $.09(10) + 4 = 4.9$
    $.9 + 4 = 4.9$
    $4.9 = 4.9$

5. $.6A + 2.8 = 4.66$
    $.6A = 4.66 - 2.8$
    $.6A = 1.86$
    $A = 1.86 \div .6$
    $A = 3.1$

6. $.6(3.1) + 2.8 = 4.66$
    $1.86 + 2.8 = 4.66$
    $4.66 = 4.66$

7. $4.6Q + .19 = 55.39$
    $4.6Q = 55.39 - .19$
    $4.6Q = 55.2$
    $Q = 55.2 \div 4.6$
    $Q = 12$

LESSON PRACTICE 22A - LESSON PRACTICE 22C

8. $4.6(12) + .19 = 55.39$
    $55.2 + .19 = 55.39$
    $55.39 = 55.39$
9. $.2M + 10.50 = 15.50$
    $.2M = 15.50 - 10.50$
    $.2M = 5.00$
    $M = 5.00 \div .2$
    $M = \$25.00$
10. $.10H + 5 = 13$
    $.10H = 13 - 5$
    $.10H = 8$
    $H = 8 \div .10 = 80 \text{ ft}$

Lesson Practice 22B

1. $1.1W + 2.1 = 5.4$
    $1.1W = 5.4 - 2.1$
    $1.1W = 3.3$
    $W = 3.3 \div 1.1$
    $W = 3$
2. $1.1(3) + 2.1 = 5.4$
    $3.3 + 2.1 = 5.4$
    $5.4 = 5.4$
3. $.8D + .2 = .76$
    $.8D = .76 - .2$
    $.8D = .56$
    $D = .56 \div .8$
    $D = .7$
4. $.8(.7) + .2 = .76$
    $.56 + .2 = .76$
    $.76 = .76$
5. $.13X + .07 = .551$
    $.13X = .551 - .07$
    $.13X = .481$
    $X = .481 \div .13$
    $X = 3.7$
6. $.13(3.7) + .07 = .551$
    $.481 + .07 = .551$
    $.551 = .551$

7. $2.2R + .47 = 1.79$
    $2.2R = 1.79 - .47$
    $2.2R = 1.32$
    $R = 1.32 \div 2.2$
    $R = .6$
8. $2.2(.6) + .47 = 1.79$
    $1.32 + .47 = 1.79$
    $1.79 = 1.79$
9. $.80M + 14.40 = 254.40$
    $.80M = 254.40 - 14.40$
    $.80M = 240.00$
    $M = 240.00 \div .80$
    $M = \$300.00$
10. $.10M + 10.00 = 15.00$
    $.10M = 15.00 - 10.00$
    $.10M = 5.00$
    $M = 5.00 \div .10$
    $M = \$50.00$

Lesson Practice 22C

1. $.7G + 5.1 = 5.59$
    $.7G = 5.59 - 5.1$
    $.7G = .49$
    $G = .49 \div .7$
    $G = .7$
2. $.7(.7) + 5.1 = 5.59$
    $.49 + 5.1 = 5.59$
    $5.59 = 5.59$
3. $3.4X + .01 = .622$
    $3.4X = .622 - .01$
    $3.4X = .612$
    $X = .612 \div 3.4$
    $X = .18$
4. $3.4(.18) + .01 = .622$
    $.612 + .01 = .622$
    $.622 = .622$
5. $8.7Y + 1.1 = 5.45$
    $8.7Y = 5.45 - 1.1$
    $8.7Y = 4.35$
    $Y = 4.35 \div 8.7$
    $Y = .5$

LESSON PRACTICE 22C - SYSTEMATIC REVIEW 22D

6. $8.7(.5)+1.1 = 5.45$
   $4.35+1.1 = 5.45$
   $5.45 = 5.45$
7. $.01B+3.3 = 3.341$
   $.01B = 3.341-3.3$
   $.01B = .041$
   $B = .041 \div .01$
   $B = 4.1$
8. $.01(4.1)+3.3 = 3.341$
   $.041+3.3 = 3.341$
   $3.341 = 3.341$
9. $.5P+1 = 4$
   $.5P = 4-1$
   $.5P = 3$
   $P = 3 \div .5$
   $P = 6$ gal for job
   $6-4 = 2$ gal to buy
10. $2.1R+.6 = 32.1$
    $2.1R = 32.1-.6$
    $2.1R = 31.5$
    $R = 31.5 \div 2.1$
    $R = 15"$

Systematic Review 22D

1. $6.8X+.9 = 22.66$
   $6.8X = 22.66-.9$
   $6.8X = 21.76$
   $X = 21.76 \div 6.8$
   $X = 3.2$
2. $6.8(3.2)+.9 = 22.66$
   $21.76+.9 = 22.66$
   $22.66 = 22.66$
3. $5.1Q+4 = 4.306$
   $5.1Q = 4.306-4$
   $5.1Q = .306$
   $Q = .306 \div 5.1$
   $Q = .06$
4. $5.1(.06)+4 = 4.306$
   $.306+4 = 4.306$
   $4.306 = 4.306$

5. $8.908 \approx 8.91$
   $6 \overline{)53.450}$
   $\underline{48000}$
   5450
   $\underline{5400}$
   50
   $\underline{48}$
6. $.271 \approx .27$
   $7 \overline{)1.900}$
   $\underline{1400}$
   500
   $\underline{490}$
   10
   $\underline{7}$
7. $.204 \approx .20$
   $14 \overline{)2.860}$
   $\underline{2800}$
   60
   $\underline{56}$
8. .382 hm
9. .00017 L
10. 5 dkg
11. done
12. $10 \times 10 \times 10 = 1,000$ m^3
13. $1.2 \times 1.2 \times 1.2 = 1.728$ ft^3
14. $1.7+.8+1.7+.8 = 5$ m
15. $11+3+11+3 = 28$ in
16. $3.14(6) = 18.84$ ft
17. $\$16.64 \times 1.15 = \$19.136$
    ($\$19.14$ rounded)
18. $.5L+12 = 132$
    $.5L = 132-12$
    $.5L = 120$
    $L = 120 \div .5$
    $L = 240$ ft^2

Systematic Review 22E

1. $.4T + 3.6 = 12$
   $.4T = 12 - 3.6$
   $.4T = 8.4$
   $T = 8.4 \div .4$
   $T = 21$

2. $.4(21) + 3.6 = 12$
   $8.4 + 3.6 = 12$
   $12 = 12$

3. $9.9W + .7 = 5.749$
   $9.9W = 5.749 - .7$
   $9.9W = 5.049$
   $W = 5.049 \div 9.9$
   $W = .51$

4. $9.9(.51) + .7 = 5.749$
   $5.049 + .7 = 5.749$
   $5.749 = 5.749$

5. 1.833 or $1.8\overline{3}$
   ```
       1.833
   24|44.000
      24 000
      19 200
         800
         720
          80
          72
           4
   ```

6. 351.66 or $351.\overline{6}$
   ```
        351.66
   6|2110.00
     1800 00
      310 00
      300 00
       10 00
          600
          400
          360
           40
           36
            4
   ```

7. 9.44 or $9.\overline{4}$
   ```
       9.44
   9|85.00
     81 00
      4 00
      3 60
        40
        36
         4
   ```

8. $.75$ kg

9. 24.5 cm

10. 60 hl

11. $.03 \times .03 \times .03 = .000027$ ft^3

12. $8 \times 8 \times 8 = 512$ m^3

13. $6.4 \times 6.4 \times 6.4 = 262.144$ in^3

14. $1.7 \times .8 = 1.36$ m^2

15. $11 \times 2 = 22$ in^2

16. $3.14(3)^2 = 28.26$ ft^2

17. $\$48.00 \times .55 = \$26.40$
    $\$48.00 - \$26.40 = \$21.60$
    $\$21.60 \times 1.06 = \$22.90$

18. $50,000$ g $= 50$ kg
    $50 \times 2.2 = 110$ lb

19. $5/9 \times 54$ games
    $54 \div 9 = 6$
    $6 \times 5 = 30$ games won

20. $.7M + 5 = 26.49$
    $.7M = 26.49 - 5.00$
    $.7M = 21.49$
    $M = 21.49 \div 7 = \$30.70$ needed
    $\$30.70 - \$26.49 = \$4.21$ still to go

Systematic Review 22F

1. $.08G + .59 = 1.39$
   $.08G = 1.39 - .59$
   $.08G = .8$
   $G = .8 \div .08$
   $G = 10$

2. $.08(10) + .59 = 1.39$
   $.8 + .59 = 1.39$
   $1.39 = 1.39$

3. $.3X + 2.4 = 2.58$
   $.3X = 2.58 - 2.4$
   $.3X = .18$
   $X = .18 \div .3$
   $X = .6$

4. $.3(.6) + 2.4 = 2.58$
   $.18 + 2.4 = 2.58$
   $2.58 = 2.58$

SYSTEMATIC REVIEW 22F - LESSON PRACTICE 23C

5. $.72\frac{6}{7}$

   $7\overline{)5.10}$
   $\underline{4\ 90}$
   $\ \ \ 20$
   $\ \ \ \underline{14}$

6. $6.27\frac{1}{2}$

   $2\overline{)12.55}$
   $\underline{12\ 00}$
   $\ \ \ \ 55$
   $\ \ \ \ \underline{40}$
   $\ \ \ \ 15$
   $\ \ \ \ \underline{14}$

7. $1.56\frac{2}{8} = 1.56\frac{1}{4}$

   $8\overline{)12.50}$
   $\underline{8\ 00}$
   $4\ 50$
   $\underline{4\ 00}$
   $\ \ \ 50$
   $\ \ \ \underline{48}$

8. 6.5 g
9. .321 km
10. 62 ml
11. $\frac{5}{2} \times \frac{5}{2} \times \frac{5}{2} = \frac{125}{8} = 15\frac{5}{8}$ ft^3
12. $.9 \times .9 \times .9 = .729$ m^3
13. $4.8 \times 4.8 \times 4.8 = 110.592$ in^3
14. $1/2(6)(16) = 48$ ft^2
15. $10 + 10 + 16 = 36$ ft
16. $1/2(3)(5.1) = 7.65$ m^2
17. $4.5 + 5.1 + 3.3 = 12.9$ m
18. $3 \times 3 = 9$ ft^2
19. $12 \times 15 = 180$ ft^2
    $180 \div 9 = 20$ yd^2
20. $20 \times \$9.99 = \$199.80$

Lesson Practice 23A

1. done
2. $6 \div 7 = .85\frac{5}{7} = 85\frac{5}{7}\%$
3. $3 \div 10 = .30 = 30\%$
4. $2 \div 5 = .40 = 40\%$
5. $7 \div 8 = .87\frac{1}{2} = 87\frac{1}{2}\%$
6. $10 \div 11 = .90\frac{10}{11} = 90\frac{10}{11}\%$
7. $\frac{13}{15} = 13 \div 15 = .86\frac{2}{3} = 86\frac{2}{3}\%$
8. $\frac{1}{6} = 1 \div 6 = .16\frac{2}{3} = 16\frac{2}{3}\%$
9. $\frac{16}{50} = 16 \div 50 = .32 = 32\%$
10. $50 - 16 = 34$ not wearing hats

    $\frac{34}{50} = 34 \div 50 = .68 = 68\%$

    (or $100\% - 32\% = 68\%$)

Lesson Practice 23B

1. $7 \div 10 = .70 = 70\%$
2. $1 \div 2 = .50 = 50\%$
3. $4 \div 9 = .44\frac{4}{9} = 44\frac{4}{9}\%$
4. $3 \div 11 = .27\frac{3}{11} = 27\frac{3}{11}\%$
5. $3 \div 5 = .60 = 60\%$
6. $1 \div 12 = .08\frac{1}{3} = 8\frac{1}{3}\%$
7. $3 \div 4 = .75 = 75\%$
8. $5 \div 6 = .83\frac{1}{3} = 83\frac{1}{3}\%$
9. $25 - 3 = 22$ correct

    $22 \div 25 = .88 = 88\%$
10. $35 \div 100 = .35 = 35\%$

Lesson Practice 23C

1. $1 \div 7 = .14\frac{2}{7} = 14\frac{2}{7}\%$
2. $2 \div 9 = .22\frac{2}{9} = 22\frac{2}{9}\%$

LESSON PRACTICE 23C - SYSTEMATIC REVIEW 23E

3. $4 \div 5 = .80 = 80\%$
4. $5 \div 12 = .41\frac{2}{3} = 41\frac{2}{3}\%$
5. $2 \div 3 = .66\frac{2}{3} = 66\frac{2}{3}\%$
6. $3 \div 8 = .37\frac{1}{2} = 37\frac{1}{2}\%$
7. $87 \div 100 = .87 = 87\%$
8. $5 \div 8 = .62\frac{1}{2} = 62\frac{1}{2}\%$
9. $1 \div 3 = .33\frac{1}{3} = 33\frac{1}{3}\%$
10. $1 \div 2 = .50 = 50\%$

Systematic Review 23D

1. $1 \div 4 = .25 = 25\%$
2. $7 \div 9 = .77\frac{7}{9} = 77\frac{7}{9}\%$
3. $5 \div 16 = .31\frac{1}{4} = 31\frac{1}{4}\%$
4. $1 \div 8 = .12\frac{1}{2} = 12\frac{1}{2}\%$
5. $.17X + 1.2 = 1.234$
   $.17X = 1.234 - 1.2$
   $.17X = .034$
   $X = .034 \div .17$
   $X = .2$
6. $.17(.2) + 1.2 = 1.234$
   $.034 + 1.2 = 1.234$
   $1.234 = 1.234$
7. $2.046 \approx 2.05$
   $13\,\overline{)26.600}$
   $\underline{26\ 000}$
   600
   $\underline{520}$
   80
   $\underline{78}$
8. $.822 \approx .82$
   $9\,\overline{)7.400}$
   $\underline{7\ 200}$
   200
   $\underline{180}$
   20
   $\underline{18}$

9. $.198 \approx .20$
   $23\,\overline{)4.570}$
   $\underline{2\ 300}$
   $2\ 270$
   $\underline{2\ 070}$
   200
   $\underline{184}$
10. done
11. $12 \times 4 \times 4.1 = 196.8 \text{ m}^3$
12. $2.3 \times 1.8 \times 3.2 = 13.248 \text{ ft}^3$
13. $38 \div 45 = .84\frac{4}{9} = 84\frac{4}{9}\%$
14. $6 \div 8 = .75 = 75\%$
15. $4 \times 2 \times 2 = 16 \text{ mi}^3$
16. $12 \times 12 = 144 \text{ in}^2$
17. $10 \times 10 = 100 \text{ ft}^2$
    $100 \times 144 = 14{,}400 \text{ in}^2$
18. $\$15.99 \times .40 = \$6.396$ or
    $\$6.40$ off the regular price

Systematic Review 23E

1. $1 \div 9 = .11\frac{1}{9} = 11\frac{1}{9}\%$
2. $2 \div 3 = .66\frac{2}{3} = 66\frac{2}{3}\%$
3. $11 \div 14 = .78\frac{1}{4} = 78\frac{1}{4}\%$
4. $4 \div 7 = .57\frac{1}{7} = 57\frac{1}{7}\%$
5. $.3Y + .09 = 1.44$
   $.3Y = 1.44 - .09$
   $.3Y = 1.35$
   $Y = 1.35 \div .3$
   $Y = 4.5$
6. $.3(4.5) + .09 = 1.44$
   $1.35 + .09 = 1.44$
   $1.44 = 1.44$

SYSTEMATIC REVIEW 23E - SYSTEMATIC REVIEW 23F

7. 8.909 or 8.9$\overline{0}$
    $11\,\overline{)98.000}$
     88 000
     10 000
     9 900
     100
     99

8. 12.9833 or 12.98$\overline{3}$
    $6\,\overline{)77.9000}$
     60 0000
     17 0000
     12 0000
     5 0000
     5000
     4800
     200
     180

9. 104.166 or 104.1$\overline{6}$
    $48\,\overline{)5000.000}$
     4800 000
     200 000
     192 000
     8 000
     4 800
     3 200
     2 880
     320
     288

10. $7 \times 6 \times 5 = 210$ in^3
11. $7.5 \times 2.4 \times 2 = 36$ m^3
12. $\dfrac{7}{2} \times \dfrac{9}{4} \times \dfrac{29}{8} =$
    $\dfrac{1827}{64} = 28\dfrac{35}{64}$ ft^3
13. $102 + 60 = 162$ total games
    $102 \div 162 = .62\dfrac{26}{27} = 62\dfrac{26}{27}\%$
14. $60 \div 162 = .37\dfrac{1}{27} = 37\dfrac{1}{27}\%$
    $.37\dfrac{1}{27} + 62\dfrac{26}{27} = .99\dfrac{27}{27} = 100\%$

15. $\$47.50 \times 1.08 = \$51.30$
16. round pizza:
    $3.14(6)^2 \approx 113$ in^2
    $\$5.65 \div 113 = \$0.05$ per in^2
    rectangular pizza:
    $12 \times 24 = 288$ in^2
    $\$11.52 \div 288 = \$0.04$ per in^2
    Rectangular pizza has better price.
17. $288 \div 144 = 2$ ft^2
18. $\$4.20 \times .75 = \$3.15$
19. $36 + 42 + 36 + 42 = 156$ in
20. $156 \div 12 = 13$ ft

Systematic Review 23F

1. $7 \div 15 = .46\dfrac{2}{3} = 46\dfrac{2}{3}\%$
2. $3 \div 7 = .42\dfrac{6}{7} = 42\dfrac{6}{7}\%$
3. $5 \div 9 = .55\dfrac{5}{9} = 55\dfrac{5}{9}\%$
4. $1 \div 10 = .10 = 10\%$
5. $6.7R + .4 = 6.43$
    $6.7R = 6.43 - .4$
    $6.7R = 6.03$
    $R = 6.03 \div 6.7$
    $R = .9$
6. $6.7(.9) + .4 = 6.43$
    $6.03 + .4 = 6.43$
    $6.43 = 6.43$
7. 1.56 $\dfrac{2}{33}$
    $33\,\overline{)51.50}$
     3300
     1850
     1650
     200
     198
     2

8. $8.23\dfrac{6}{8} = 8.23\dfrac{3}{4}$

    $8\overline{)65.90}$
    $\underline{6400}$
    190
    $\underline{160}$
    30
    $\underline{24}$
    6

9. $11.96\dfrac{4}{6} = 11.96\dfrac{2}{3}$

    $06\overline{)71.80}$
    $\underline{6000}$
    1180
    $\underline{600}$
    580
    $\underline{540}$
    40
    $\underline{36}$
    4

10. $80 \times 90 \times 60 = 432{,}000$ in^3
11. $13.1 \times 4.6 \times 2.9 = 174.754$ m^3
12. $.09 \times .05 \times 1 = .0045$ ft^3
13. $3 \times 3 \times 3 = 27$ ft^3
14. $27 \times 2\dfrac{1}{3} = \dfrac{\cancel{27}^9}{1} \times \dfrac{7}{\cancel{3}} = \dfrac{63}{1} = 63$ ft^3
15. $6 \times 27 = 162$ ft^3
16. $6 \times \$55.50 = \$333$
17. $6 \div 8 = .75 = 75\%$
18. $34 \div 8 = 4.25$, so 5 trips
19. $\$32 \times .30 = \$9.60$
    $\$32 - \$9.60 = \$22.40$ before tax
    $\$22.40 \times 1.04 = \$23.296$ or $\$23.30$
    $\$25 - \$23.30 = \$1.70$
20. $7.8H = 27.3$
    $\dfrac{\cancel{7.8}}{\cancel{7.8}}H = \dfrac{27.3}{7.8}$
    $27.3 \div 7.8 = 3.5$ ft

Lesson Practice 24A

1. done
2. $\dfrac{96}{100} = \dfrac{24}{25}$
3. $\dfrac{982}{1000} = \dfrac{491}{500}$
4. $\dfrac{6}{1000} = \dfrac{3}{500}$
5. $\dfrac{18}{1000} = \dfrac{9}{500}$
6. $\dfrac{9}{10}$
7. $\dfrac{885}{1000} = \dfrac{177}{200}$
8. $\dfrac{84}{100} = \dfrac{21}{25}$
9. $\dfrac{32}{100} = \dfrac{8}{25}$
10. $\dfrac{15}{1000} = \dfrac{3}{200}$
11. $\dfrac{20}{100} = \dfrac{1}{5}$; 5 pieces
12. $\dfrac{45}{100} = \dfrac{9}{20}$; 9 people

Lesson Practice 24B

1. $\dfrac{915}{1000} = \dfrac{183}{200}$
2. $\dfrac{53}{1000}$
3. $\dfrac{7}{1000}$
4. $\dfrac{62}{100} = \dfrac{31}{50}$
5. $\dfrac{75}{100} = \dfrac{3}{4}$
6. $\dfrac{52}{100} = \dfrac{13}{25}$
7. $\dfrac{28}{1000} = \dfrac{7}{250}$
8. $\dfrac{725}{1000} = \dfrac{29}{40}$
9. $\dfrac{6}{10} = \dfrac{3}{5}$
10. $\dfrac{95}{100} = \dfrac{19}{20}$

LESSON PRACTICE 24B - SYSTEMATIC REVIEW 24D

11. $\frac{4}{10} = \frac{2}{5}$; 2 juncos
12. $\frac{125}{1000} = \frac{1}{8}$; 8 pieces

Lesson Practice 24C

1. $\frac{68}{100} = \frac{17}{25}$
2. $\frac{25}{1000} = \frac{1}{40}$
3. $\frac{84}{100} = \frac{21}{25}$
4. $\frac{16}{1000} = \frac{2}{125}$
5. $\frac{8}{1000} = \frac{1}{125}$
6. $\frac{7}{10}$
7. $\frac{325}{1000} = \frac{13}{40}$
8. $\frac{743}{1000}$
9. $\frac{56}{100} = \frac{14}{25}$
10. $\frac{11}{100}$
11. $\frac{4}{100} = \frac{1}{25}$; 25 pieces
12. $.75 \times 4 = 3$ houses

Systematic Review 24D

1. $\frac{25}{1000} = \frac{1}{40}$
2. $\frac{18}{100} = \frac{9}{50}$
3. $1 \div 5 = .20 = 20\%$
4. $5 \div 8 = .62\frac{1}{2} = 62\frac{1}{2}\%$
5. $.19X + .73 = .825$
   $.19X = .825 - .73$
   $.19X = .095$
   $X = .095 \div .19$
   $X = .5$

6. $.19(.5) + .73 = .825$
   $.095 + .73 = .825$
   $.825 = .825$

7. $\phantom{15\overline{)}}2.306 = 2.31$
   $15\overline{)34.600}$
   $\phantom{15\overline{)}}\underline{30000}$
   $\phantom{15\overline{)}}4600$
   $\phantom{15\overline{)}}\underline{4500}$
   $\phantom{15\overline{)}}100$
   $\phantom{15\overline{)}}\underline{90}$
   $\phantom{15\overline{)}}10$

8. $\phantom{8.\overline{)}}.837 = .84$
   $8.\overline{)6.700}$
   $\phantom{8.\overline{)}}\underline{6400}$
   $\phantom{8.\overline{)}}300$
   $\phantom{8.\overline{)}}\underline{240}$
   $\phantom{8.\overline{)}}60$
   $\phantom{8.\overline{)}}\underline{56}$
   $\phantom{8.\overline{)}}4$

9. $\phantom{34.\overline{)}}.036 = .04$
   $34.\overline{)1.230}$
   $\phantom{34.\overline{)}}\underline{1020}$
   $\phantom{34.\overline{)}}210$
   $\phantom{34.\overline{)}}\underline{204}$
   $\phantom{34.\overline{)}}6$

10. $2 + 7 + 9 = 18$
    $18 \div 3 = 6$
11. $5 + 5 + 9 + 13 = 32$
    $32 \div 4 = 8$
12. $2 + 5 + 8 + 9 = 24$
    $24 \div 4 = 6$
13. $77 + 80 + 95 + 100 = 352$
    $352 \div 4 = 88$
14. $19.06 \times 60 = 1{,}143.6$ ft
15. $1.15 \times 5 = 5.75$ hr
16. $\frac{1}{2}(.24 \times .5) = .06$ m^2
17. $2{,}500 \div 1{,}000 = 2.5$ liters
18. $4 \times 10 = 40$ hm

Systematic Review 24E

1. $\dfrac{625}{1000} = \dfrac{5}{8}$
2. $\dfrac{21}{100}$
3. $2 \div 7 = .28\dfrac{4}{7} = 28\dfrac{4}{7}\%$
4. $1 \div 4 = .25 = 25\%$
5. $.3X + 9.1 = 9.244$
   $.3X = 9.244 - 9.1$
   $.3X = .144$
   $X = .144 \div .3$
   $X = .48$
6. $.3(.48) + 9.1 = 9.244$
   $.144 + 9.1 = 9.244$
   $9.244 = 9.244$
7. $9.044 = 9.0\overline{4}$
   $9\overline{)81.400}$
   $\underline{81000}$
   400
   $\underline{360}$
   40
   $\underline{36}$
   4
8. $48.66 = 48.\overline{6}$
   $3\overline{)146.00}$
   $\underline{12000}$
   2600
   $\underline{2400}$
   200
   $\underline{180}$
   20
   $\underline{18}$
   2
9. $4.8181 = 4.\overline{81}$
   $11\overline{)53.0000}$
   $\underline{440000}$
   90000
   $\underline{88000}$
   2000
   $\underline{1100}$
   900
   $\underline{880}$
   20
10. $4 + 5 + 6 = 15$
    $15 \div 3 = 5$
11. $5 + 8 + 8 = 21$
    $21 \div 3 = 7$
12. $6 + 7 + 9 + 10 = 32$
    $32 \div 4 = 8$
13. $76 + 81 + 89 + 92 = 338$
    $338 \div 4 = 84.5$
14. $10 \times 10 \times 10 = 1{,}000 \text{ m}^3$
15. $6 + 9 + 11 = 26 \text{ in}$
16. $10{,}000{,}000 \div 1{,}000 = 10{,}000 \text{ km}$
17. $384 \times .75 = 288 \text{ oz}$
18. $288 \div 16 = 18 \text{ lb}$
19. $\$9.00 \times 1.10 = \$9.90$
20. $3.14(4)^2 = 50.24 \text{ in}^2$

Systematic Review 24F

1. $\dfrac{92}{100} = \dfrac{23}{25}$
2. $\dfrac{45}{1000} = \dfrac{9}{200}$
3. $1 \div 9 = .11\dfrac{1}{9} = 11\dfrac{1}{9}\%$
4. $3 \div 5 = .60 = 60\%$
5. $.2X + 75 = 75.18$
   $.2X = 75.18 - 75$
   $.2X = .18$
   $X = .18 \div .2$
   $X = .9$
6. $.2(.9) + 75 = 75.18$
   $.18 + 75 = 75.18$
   $75.18 = 75.18$

7. $166.66\frac{4}{6} = 166.66\frac{2}{3}$

   $6\overline{)1000.00}$
   $\underline{60000}$
   40000
   $\underline{36000}$
   4000
   $\underline{3600}$
   400
   $\underline{360}$
   40
   $\underline{36}$
   4

8. $.32\frac{1}{2}$

   $2\overline{)0.65}$
   $\underline{60}$
   5
   $\underline{4}$
   1

9. $2.11\frac{3}{17}$

   $17\overline{)35.90}$
   $\underline{3400}$
   190
   $\underline{170}$
   20
   $\underline{17}$
   3

10. $3 + 6 + 9 = 18$
    $18 \div 3 = 6$
11. $4 + 8 + 10 + 14 = 36$
    $36 \div 4 = 9$
12. $1 + 3 + 6 + 10 = 20$
    $20 \div 4 = 5$
13. $21 + 34 + 21 + 34 = 110$ in
14. $(2)(3.14)(2) = 25.12$ ft
15. $\$2,500 \times 1.17 = \$2,925$
16. $250 \div 100 = 2.5$, so he must buy 3 packs
17. $\$1.29 \times 3 = \$3.87$
    $\$3.87 \times 1.05 \approx \$4.06$
18. $3.25 \times 40 = 130$ lb
19. $130 \div 10 = 13$ bags
20. $13 \times \$1.45 = \$18.85$

Lesson Practice 25A

1. done
2. $\frac{3+8+8+8+9+9+11}{7} = 8$
   mean = 8
   median = 8
   mode = 8
3. $\frac{3+8+9+9+11}{5} = 8$
   mean = 8
   median = 9
   mode = 9
4. $\frac{3+4+6+6+6+8+9}{7} = 6$
   mean = 6
   median = 6
   mode = 6
5. $\frac{3+8+8+10+12+19}{6} = 10$
   mean = 10
   median = $(8+10)/2 = 9$
   mode = 8
6. $\frac{11+12+15+15+33+34}{6} = 20$
   mean = 20
   median = 15
   mode = 15
7. 6
8. 29° was the average temperature
9. 3

Lesson Practice 25B

1. $\frac{10+11+15+15+15+16+23}{7} = 15$
   mean = 15
   median = 15
   mode = 15
2. $\frac{1+1+1+1+2+2+2+3+3+4}{10} = 2$
   mean = 2
   median = 2
   mode = 1

3. $\dfrac{3+4+7+7+7+9+19}{7} = 8$
   mean = 8
   median = 7
   mode = 7

4. $\dfrac{13+14+24+24+25}{5} = 20$
   mean = 20
   median = 24
   mode = 24

5. $\dfrac{15+17+17+17+19+23}{6} = 18$
   mean = 18
   median = 17
   mode = 17

6. $\dfrac{9+10+10+10+10+11+13+13+13}{9} = 1$
   mean = 11
   median = 10
   mode = 10

7. $17+30+32+35+40 = 154$
   $154 \div 5 = 30.8°$

8. 98

9. median

Lesson Practice 25C

1. $\dfrac{5+8+8+9+15}{5} = 9$
   mean = 9
   median = 8
   mode = 8

2. $\dfrac{9+11+11+14+20+23+24}{7} = 16$
   mean = 16
   median = 14
   mode = 11

3. $\dfrac{7+7+7+7+14+18}{6} = 10$
   mean = 10
   median = 7
   mode = 7

4. $\dfrac{1+6+6+19}{4} = 8$
   mean = 8
   median = 6
   mode = 6

5. $\dfrac{11+13+13+13+19+21}{6} = 15$
   mean = 15
   median = 13
   mode = 13

6. $\dfrac{8+9+11+14+15+15+19}{7} = 13$
   mean = 13
   median = 14
   mode = 15

7. $5+6+7+8+9 = 35$
   $35 \div 5 = 7$

8. 5'10"; median

9. 8; mode

Systematic Review 25D

1. $\dfrac{5+6+6+9+10+10+10}{7} = 8$
   mean = 8
   median = 9
   mode = 10

2. $\dfrac{7+8+11+11+13}{5} = 10$
   mean = 10
   median = 11
   mode = 11

3. $\dfrac{13}{100}$

4. $\dfrac{350}{1000} = \dfrac{35}{100} = \dfrac{7}{20}$

5. $2 \div 9 = .22\dfrac{2}{9} = 22\dfrac{2}{9}\%$

6. $1 \div 10 = .10 = 10\%$

7. $.7X + 1.5 = 3.88$
   $.7X = 3.88 - 1.5$
   $.7X = 2.38$
   $X = 2.38 \div .7$
   $X = 3.4$

8. $.7(3.4) + 1.5 = 3.88$
    $2.38 + 1.5 = 3.88$
    $3.88 = 3.88$
9. $3.14(4.5)^2 = 63.59$ in^2 (rounded)
10. $3.14(9) = 28.26$ in
11. $8 \times 8 = 64$
12. $3 \times 3 = 9$
13. $10 \times 10 = 100$
14. $15 + 15 + 24 = 54$ ft
15. $12 + 16 + 20 = 48$ m
16. $1.5 + 2.5 + 2 = 6$ ft
17. $\frac{1}{2}(24 \times 9) = 108$ ft^2
18. $\frac{1}{2}(12 \times 16) = 96$ m^2
19. $\frac{1}{2}(1.2 \times 2.5) = 1.5$ ft^2
20. $50 \times 1,000 = 50,000$ mg total
    $50,000 \div 100 = 500$ mg per tablet

Systematic Review 25E

1. $\frac{2+4+4+4+7+9}{6} = 5$
    mean = 5
    median = 4
    mode = 4
2. $\frac{3+8+8+15+16+29+33}{7} = 16$
    mean = 16
    median = 15
    mode = 8
3. $\frac{8}{10} = \frac{4}{5}$
4. $\frac{48}{100} = \frac{12}{25}$
5. $5 \div 6 = .83\frac{1}{3} = 83\frac{1}{3}\%$
6. $3 \div 15 = .2 = 20\%$
7. $1.1X + 2.2 = 9.9$
    $1.1X = 9.9 - 2.2$
    $1.1X = 7.7$
    $X = 7.7 \div 1.1$
    $X = 7$

8. $1.1(7) + 2.2 = 9.9$
    $7.7 + 2.2 = 9.9$
    $9.9 = 9.9$
9. $2.14(5)^2 = 78.5$ in^2
10. $3.14(10) = 31.4$ in
11. $2 \times 2 \times 2 = 8$
12. $1 \times 1 \times 1 \times 1 \times 1 = 1$
13. $7 \times 7 = 49$
14. $2.6 + 2.6 + 2.6 + 2.6 = 10.4$ ft
15. $.1 + .51 + .1 + .51 = 1.22$ m
16. $3.1 + 5.4 + 3.1 + 5.4 = 17$ ft
17. $2.6 \times 2.6 = 6.76$ ft^2
18. $.1 \times .51 = .051$ m^2
19. $3.1 \times 5.4 = 16.74$ ft^2
20. $30,300 \div 1,000 = 30.3$ kl

Systematic Review 25F

1. $\frac{1+1+1+1+1+5+6+8}{8} = 3$
    mean = 3
    median = 1
    mode = 1
2. $\frac{5+9+9+9+13+15}{6} = 10$
    mean = 10
    median = 9
    mode = 9
3. $\frac{8}{100} = \frac{2}{25}$
4. $\frac{175}{1000} = \frac{7}{40}$
5. $9 \div 10 = .9 = 90\%$
6. $7 \div 8 = .87\frac{1}{2} = 87\frac{1}{2}\%$
7. $8X + .09 = 1.69$
    $8X = 1.69 - .09$
    $8X = 1.6$
    $X = 1.6 \div 8$
    $X = .2$
8. $8(.2) + .09 = 1.69$
    $1.6 + .09 = 1.69$
    $1.69 = 1.69$

9. $3.14(3.1)^2 = 30.1754 \text{ ft}^2$
10. $3.14(6.2) = 19.468 \text{ ft}$
11. $4 \times 4 = 16$
12. $6 \times 6 = 36$
13. $5 \times 5 \times 5 = 125$
14. $4.2 + 5 + 4.2 + 5 = 18.4 \text{ m}$
15. $18 + 9 + 18 + 9 = 54 \text{ mm}$
16. $2.4 + 1.8 + 2.4 + 1.8 = 8.4 \text{ in}$
17. $4.2 \times 4 = 16.8 \text{ m}^2$
18. $18 \times 6 = 108 \text{ mm}^2$
19. $2.4 \times 1.2 = 2.88 \text{ in}^2$
20. $3 \times 10 = 30 \text{ pieces}$

Lesson Practice 26A

1. done
2. done
3. done
4. $\frac{7}{50}$
5. $\frac{20}{300} = \frac{1}{15}$
6. $\frac{240}{300} = \frac{4}{5}$
7. $\frac{40}{300} = \frac{2}{15}$
8. $\frac{300-20}{300} = \frac{280}{300} = \frac{14}{15}$
9. $\frac{2}{1000} = \frac{1}{500}$
10. $\frac{10}{1000} = \frac{1}{100}$
11. $\frac{5}{500} = \frac{1}{100}$
12. $\frac{32}{192+32} = \frac{32}{224} = \frac{1}{7}$

Lesson Practice 26B

1. $\frac{25}{80} = \frac{5}{16}$
2. $\frac{20+15}{80} = \frac{35}{80} = \frac{7}{16}$
3. $\frac{20+25+15}{80} = \frac{60}{80} = \frac{3}{4}$
4. $\frac{20+25+13}{80} = \frac{58}{80} = \frac{29}{40}$
5. $\frac{10}{120} = \frac{1}{12}$
6. $\frac{100}{120} = \frac{5}{6}$
7. $\frac{10}{120} = \frac{1}{12}$
8. $\frac{100+10}{120} = \frac{110}{120} = \frac{11}{12}$
9. $\frac{4}{10} = \frac{2}{5}$
10. $\frac{7}{10}$
11. $\frac{4}{10} = \frac{2}{5}$
12. $\frac{3}{10}$

Lesson Practice 26C

1. $\frac{30}{250} = \frac{3}{25}$
2. $\frac{50+40+30+30}{250} = \frac{150}{250} = \frac{3}{5}$
3. $\frac{40+30}{250} = \frac{70}{250} = \frac{7}{25}$
4. $\frac{100+50+30}{250} = \frac{180}{250} = \frac{18}{25}$
5. $\frac{10}{3500} = \frac{1}{350}$
6. $\frac{500}{3500} = \frac{1}{7}$
7. $\frac{3500-500}{3500} = \frac{3000}{3500} = \frac{6}{7}$
8. $\frac{4}{10} = \frac{2}{5}$
9. $\frac{7}{10}$
10. $\frac{1}{10}$
11. $\frac{2}{10} = \frac{1}{5}$
12. $\frac{568 \div 2}{568} = \frac{284}{568} = \frac{1}{2}$

Systematic Review 26D

1. $\dfrac{60}{540} = \dfrac{1}{9}$
2. $\dfrac{450}{540} = \dfrac{5}{6}$
3. $\dfrac{450+60}{540} = \dfrac{510}{540} = \dfrac{17}{18}$
4. $\dfrac{5+9+9+9+13+15}{6} = 10$
   mean = 10
   median = 9
   mode = 9
5. $\dfrac{1+1+3+3+3+3+5+6+6+9}{10} = 4$
   mean = 4
   median = 3
   mode = 3
6. $\dfrac{34}{100} = \dfrac{17}{50}$
7. $\dfrac{178}{1000} = \dfrac{89}{500}$
8. $.75 = 75\%$
9. $.55\dfrac{5}{9} = 55\dfrac{5}{9}\%$
10. $5+4=9;\ 9\times8=72;\ 72-2=70;$
    $70\div10=7;\ 7+3=10$
11. $7-3=4;\ 4\times6=24;\ 24\div3=8;$
    $8\times9=72$
12. $\dfrac{2}{5000} = \dfrac{1}{2500}$
13. $45 \div .9 = 50$ customers
14. $\$48.00 \times .20 = \$9.60$
    $\$48.00 - \$9.60 = \$38.40$

Systematic Review 26E

1. $\dfrac{100}{500} = \dfrac{1}{5}$
2. $\dfrac{250+100}{500} = \dfrac{350}{500} = \dfrac{7}{10}$
3. $\dfrac{30+40+100}{500} = \dfrac{170}{500} = \dfrac{17}{50}$
4. $\dfrac{7+8+10+10+14+17}{6} = 11$
   mean = 11
   median = 10
   mode = 10
5. $\dfrac{15+21+22+27+27}{5} = 22.4$
   mean = 22.4
   median = 22
   mode = 27
6. $\dfrac{48}{100} = \dfrac{12}{25}$
7. $\dfrac{525}{1000} = \dfrac{21}{40}$
8. $.83\dfrac{1}{3} = 83\dfrac{1}{3}\%$
9. $.91\dfrac{2}{3} = 91\dfrac{2}{3}\%$
10. $36{,}320 \div 1{,}000 = 36.32$ kg
11. $3.14(10)^2 = 314$ ft^2
12. $25.02 + 20.6 + .25 = 45.87$ gal
13. $125.5 \times 200 = 25{,}100$ ft^2
14. $9+4=13;\ 13+3=16;\ 16+4=20;$
    $20\div5=4;\ 4-0=4$
15. $10\div2=5;\ 5+2=7;\ 7\times7=49;$
    $49+1=50;\ 50-20=30$

Systematic Review 26F

1. $\dfrac{2}{10} = \dfrac{1}{5}$
2. $\dfrac{9}{10}$
3. $\dfrac{4}{10} = \dfrac{2}{5}$
4. $\dfrac{17+17+19+22+25}{5} = 20$
   mean = 20
   median = 19
   mode = 17
5. $\dfrac{1+1+1+7+7+8+9+14}{8} = 6$
   mean = 6
   median = 7
   mode = 1

6. $\dfrac{65}{100} = \dfrac{13}{20}$
7. $\dfrac{125}{1000} = \dfrac{1}{8}$
8. $.42\dfrac{6}{7} = 42\dfrac{6}{7}\%$
9. $.90 = 90\%$
10. $200 \times .87 = 174$
11. $1 \times 10 = 10$ runners
12. $3.14(12) = 37.68$; 38 lengths
13. meter
14. $\dfrac{1}{2} \times 8 = 4$; $4 + 3 = 7$;
    $7 \times 7 = 49$; $49 + 1 = 50$
15. $8 + 9 = 17$; $17 - 7 = 10$;
    $10 \div 5 = 2$; $2 \times 4 = 8$

Lesson Practice 27A

1. ray
2. line segment
3. geometry
4. point
5. line
6. d: line
7. c: ray
8. b: line segment
9. e: point
10. a: infinity
11. A
12. $\overleftrightarrow{ST}$ or $\overleftrightarrow{TS}$

Lesson Practice 27B

1. ray
2. line segment
3. endpoint
4. points
5. point
6. d
7. a
8. c
9. b
10. $\overline{LM}$ or $\overline{ML}$
11. $\overline{QR}$
12. line

Lesson Practice 27C

1. point
2. line
3. ray
4. line segment
5. ray
6. e
7. d
8. a
9. c
10. b
11. $\overleftrightarrow{DE}$ or $\overleftrightarrow{ED}$
12. $\overline{YZ}$ or $\overline{ZY}$

Systematic Review 27D

1. length; width
2. length; width
3. one
4. two
5. $\overrightarrow{EF}$
6. $3 + 5 + 5 + 5 + 6 + 7 + 7 + 9 + 16 = 63$
   mean $= 63 \div 9 = 7$
7. median $= 6$
8. mode $= 5$
9. $\dfrac{15}{100} = \dfrac{3}{20}$
10. $\dfrac{375}{1000} = \dfrac{3}{8}$
11. $.6 + .045 = .645$
12. $113 + .19 = 113.19$
13. $2.9 + .11 = 3.01$
14. 76
15. 55
16. $\dfrac{1}{12}$

17. $\dfrac{1}{8}$

18. $10 \div 2 = 5$; $5 \times 4 = 20$;
    $20 \div 2 = 10$; $10 + 7 = 17$

Systematic Review 27E
1. line; ray
2. point
3. line segment
4. endpoint
5. $\overleftrightarrow{GK}$ or $\overleftrightarrow{KG}$
6. $1+1+1+7+7+8+9+14 = 48$
   mean = $48 \div 8 = 6$
7. median = 7
8. mode = 1
9. $\dfrac{65}{100} = \dfrac{13}{20}$
10. $\dfrac{1}{1000}$
11. $11.4 - .6 = 10.8$
12. $35 - .42 = 34.58$
13. $5.15 - .9 = 4.25$
14. 890
15. north
16. $\dfrac{15}{150} = \dfrac{1}{10}$
17. $3.14 \times 12{,}732 = 39{,}978$ km
18. $4 \times 2 = 8$; $8 + 1 = 9$;
    $9 - 7 = 2$; $2 \div 2 = 1$

Systematic Review 27F
1. infinite
2. point
3. endpoints
4. endpoint or origin
5. $\overline{MN}$ or $\overline{NM}$
6. $13+13+13+13+14+18 = 84$
   mean = $84 \div 6 = 14$
7. median = 13
8. mode = 13

9. $\dfrac{82}{100} = \dfrac{41}{50}$
10. $\dfrac{5}{1000} = \dfrac{1}{200}$
11. $.5 \times .7 = .35$
12. $.02 \times .4 = .008$
13. $1.31 \times .28 = .3668$
14. because it goes the whole way around the city
15. $\dfrac{1}{7}$
16. $4.5 \times 1.9 \times 6 = 51.3$ m^3
17. $7 \times 1.12 = 7.84$ ft
18. $\dfrac{1}{3} \times 12 = 4$; $4 + 5 = 9$;
    $9 + 1 = 10$; $10 \times 4 = 40$

Lesson Practice 28A
1. length; width
2. two
3. plane
4. congruent
5. equal
6. similar
7. b
8. a
9. d
10. c
11. g
12. e
13. h
14. f

Lesson Practice 28B
1. equal
2. similar
3. congruent
4. plane
5. point
6. geometry
7. d

8. c
9. b
10. a
11. h
12. g
13. e
14. f

Lesson Practice 28C

1. point
2. line, line segment, ray
3. plane
4. equal
5. congruent
6. similar
7. c
8. d
9. a
10. b
11.
12. =
13. ~
14. ≅

Systematic Review 28D

1. congruent
2. length; width
3. similar
4. length; width
5. one
6. length; width
7. two
8. .7T + 1.4 = 4.9
   .7T = 4.9 − 1.4
   .7T = 3.5
   T = 3.5 ÷ .7
   T = 5
9. .7(5) + 1.4 = 4.9
   3.5 + 1.4 = 4.9
   4.9 = 4.9

10. $33\frac{1}{3}\%$
11. 50%
12. 75%
13. .15 ÷ .9 = .17
14. 1.28 ÷ 6.2 = .21
15. .105 ÷ .5 = .21
16. 3 + 5 + 6 + 6 + 10 = 30
    30 ÷ 5 = 6
17. 90
18. 6 ÷ 3 = 2; 2 × 4 = 8;
    8 − 3 = 5; 5 + 6 = 11

Systematic Review 28E

1. h
2. e
3. b
4. f
5. d
6. c
7. a
8. i
9. g
10. $66\frac{2}{3}\%$
11. 25%
12. 20%
13. $.9133 = .91\overline{3}$
    $3\overline{)2.7400}$
    27000
    ‾
    400
    300
    ‾
    100
    900
    ‾
    10
    9
    ‾
    1
14. $90.66 = 90.\overline{6}$
    $9.\overline{)816.00}$
    81000
    ‾
    600
    540
    ‾
    60
    54
    ‾
    6

15. $11 \overline{)45.600}$ = 4.145 = 4.1$\overline{45}$
     44000
     1600
     1100
     500
     440
     60
     55
     5

16. 251
17. 176
18. $9 \div 3 = 3$; $3 + 2 = 5$;
    $5 \times 7 = 35$; $35 - 5 = 30$

Systematic Review 28F

1. ∞
2. —
3. →
4. ↔
5. •
6. =
7. ~
8. ≅
9. ▱
10. 40%
11. $33\frac{1}{3}$%
12. 80%
13. .73 $\frac{13}{19}$
14. .85 $\frac{5}{7}$
15. .55 $\frac{5}{9}$
16. 12"
17. north; 435
18. $\frac{1}{4} \times 12 = 3$; $3 + 7 = 10$;
    $10 \div 2 = 5$; $5 \times 5 = 25$

Lesson Practice 29A

1. angles
2. rays
3. 90
4. 360
5. vertex
6. circle
7. Greek
8. done
9. ∠QXT; ∠TXQ
10. 90°
11. four

Lesson Practice 29B

1. angle
2. right
3. endpoint
4. circle
5. right angle
6. lines
7. vertex
8. ∠ZYX; ∠XYZ
9. ∠LMN; ∠NML
10. right
11. 90°

Lesson Practice 29C

1. 90
2. 360
3. vertex
4. Greek
5. 1/4
6. 4
7. rays
8. ∠CBA; ∠ABC
9. ∠YRD; ∠DRY
10. 90°
11. 90°

Systematic Review 29D

1. circle
2. 360°
3. 90°
4. similar
5. length; width
6. two
7. congruent
8. $.85M = .425$
   $M = .425 \div .85$
   $M = .5$
9. $.85(.5) = .425$
   $.425 = .425$
10. $\frac{95}{100} = \frac{19}{20}$
11. $\frac{7}{10}$
12. $\frac{685}{1000} = \frac{137}{200}$
13. $32 \div 8 = 4$
    $4 \times 3 = 12$
14. $12 \div 6 = 2$
    $2 \times 1 = 2$
15. $72 \div 9 = 8$
    $8 \times 5 = 40$
16. $\frac{950-2}{950} = \frac{948}{950} = \frac{474}{475}$
17. $8.5 + 9.5 + 10 + 10 + 12 = 50$
    $50 \div 5 = 10$ lb
18. $\frac{1}{5} \times 15 = 3$; $3 + 8 = 11$;
    $11 - 2 = 9$; $9 \times 9 = 81$

Systematic Review 29E

1. b
2. c
3. f
4. e
5. a
6. g
7. d

Systematic Review 29D (cont.)

8. $.09Q = .315$
   $Q = .315 \div .09 = 3.5$
9. $.09(3.5) = .315$
   $.315 = .315$
10. $\frac{24}{100} = \frac{6}{25}$
11. $\frac{9}{10}$
12. $\frac{278}{1000} = \frac{139}{500}$
13. $6\frac{1}{6} + 8\frac{2}{5} = 6\frac{5}{30} + 8\frac{12}{30} = 14\frac{17}{30}$
14. $3 - 2\frac{1}{3} = 2\frac{3}{3} - 2\frac{1}{3} = \frac{2}{3}$
15. $7\frac{2}{9} + 9\frac{4}{5} = 7\frac{10}{45} + 9\frac{36}{45} =$
    $16\frac{46}{45} = 17\frac{1}{45}$
16. $3,405 \div 1,000 = 3.405$ kg
17. south
18. $20 \div 10 = 2$; $2 \times 6 = 12$;
    $12 + 3 = 15$; $15 - 8 = 7$

Systematic Review 29F

1. right
2. 1/4
3. length; width
4. equal
5. line segment
6. rays
7. ∠BCA; ∠ACB
8. $3.3X + .79 = 4.42$
   $3.3X = 4.42 - .79$
   $3.3X = 3.63$
   $X = 3.63 \div 3.3$
   $X = 1.1$
9. $3.3(1.1) + .79 = 4.42$
   $3.63 + .79 = 4.42$
   $4.42 = 4.42$
10. $\frac{75}{100} = \frac{3}{4}$
11. $\frac{9}{100}$

SYSTEMATIC REVIEW 29F - SYSTEMATIC REVIEW 30D

12. $\dfrac{138}{1000} = \dfrac{69}{500}$

13. $\dfrac{\cancel{4}}{9} \times \dfrac{5}{\cancel{8}_2} = \dfrac{5}{18}$

14. $\dfrac{\cancel{2}}{\cancel{3}} \times \dfrac{\cancel{9}^3}{\cancel{10}_5} = \dfrac{3}{5}$

15. $\dfrac{3}{\cancel{24}} \times \dfrac{\cancel{2}}{5} = \dfrac{3}{10}$

16. $\$175.62 - \$39.59 = \$136.03$

17. $2.75 \times 12.25 = 33.6875$ gal

18. $8 \div 4 = 2;\ 2 \times 5 = 10;$
    $10 - 6 = 4;\ 4 + 9 = 13$

Lesson Practice 30A
1. acute
2. obtuse
3. straight
4. b
5. d
6. a
7. c
8. right
9. obtuse
10. acute
11. straight
12. straight

Lesson Practice 30B
1. straight
2. obtuse
3. acute
4. c
5. b
6. d
7. a
8. straight
9. obtuse
10. right
11. acute
12. $360° \div 6 = 60°$; acute

Lesson Practice 30C
1. 90°; 180°
2. 0°; 90°
3. 180°
4. 90°
5. b
6. c
7. a
8.
9.
10.
11.
12. 180°

Systematic Review 30D
1. acute
2. length; width
3. 90°
4. similar
5. congruent
6. two
7. vertex
8. $.07X + .52 = .527$
   $.07X = .527 - .52$
   $.07X = .007$
   $X = .1$
9. $.07(.1) + .52 = .527$
   $.007 + .52 = .527$
   $.527 = .527$
10. $5 \div 6 = .83$
11. $8 \div 13 = .62$
12. $4 \div 5 = .8$
13. $\dfrac{1}{3} \div \dfrac{4}{7} = \dfrac{1}{3} \times \dfrac{7}{4} = \dfrac{7}{12}$
14. $\dfrac{3}{10} \div \dfrac{3}{5} = \dfrac{\cancel{3}}{\cancel{10}_2} \times \dfrac{\cancel{5}}{\cancel{3}} = \dfrac{1}{2}$

15. $\dfrac{7}{8} \div \dfrac{1}{2} = \dfrac{7}{\cancel{8}_4} \times \dfrac{\cancel{2}}{1} = \dfrac{7}{4} = 1\dfrac{3}{4}$

16. $3.14(26) = 81.64$ in

17. $\dfrac{200}{200+400} = \dfrac{200}{600} = \dfrac{1}{3}$

18. $30 \div 10 = 3$
    $3 \times 4 = 12$
    $12 + 2 = 14$
    $14 - 7 = 7$

Systematic Review 30E

1. b
2. e
3. f
4. c
5. a
6. d
7. g
8. ↔
9. ∟
10. $\dfrac{725}{1000} = \dfrac{29}{40}$
11. $\dfrac{3}{10}$
12. $\dfrac{95}{100} = \dfrac{19}{20}$
13. $65 \div 1{,}000 = .065$
14. $21 \div 10 = 2.1$
15. $.04 \times 100 = 4$
16. round:
    $3.14(6^2) = 113.04$ in^2
    square:
    $10.5 \times 10.5 = 110.25$ in^2
    $113.04 > 110.25$
    round is bigger
17. $\$5.95 \times 1.03 \approx \$6.13$
    $\$10.00 - \$6.13 = \$3.87$
18. $10 \times 4 = 40$
    $40 \div 5 = 8$
    $8 \div 4 = 2$
    $2 \div 2 = 1$

Systematic Review 30F

1. acute
2. obtuse
3. length; width
4. equal
5. 1/4
6. congruent
7. ∠XTQ; ∠QTX
8. $3+5+5+5+6+7+7+9+16 = 63$
   mean $= 63 \div 9 = 7$
   median $= 6$
   mode $= 5$
9. $9+13+27+27+30+32+37 = 175$
   mean $= 175 \div 7 = 25$
   median $= 27$
   mode $= 27$
10. 350%
11. 1%
12. 45%
13. $.35 \times 200 = 70$
14. $.02 \times 16 = .32$
15. $2.50 \times 150 = 375$
16. $\dfrac{5}{695} = \dfrac{1}{139}$
17. $6^2 = 6 \times 6 = 36$
18. $8 + 4 = 12$
    $12 - 6 = 6$
    $6 + 3 = 9$
    $9 \times 2 = 18$

Test Solutions

Test 1
1. $1 \times 1 \times 1 \times 1 = 1$
2. $10 \times 10 = 100$
3. $13 \times 13 = 169$
4. $3 \times 3 \times 3 = 27$
5. $12 \times 12 = 144$
6. $7 \times 7 = 49$
7. $5^2 = 25$
8. $4^3 = 64$
9. $2^4 = 16$
10. $6 \times 6 \times 6 \times 6 = 6^4$
11. $15 \times 15 = 15^2$
12. $9 \times 9 \times 9 = 9^3$
13. $8 \times 8 \times 8 = 8^3$
14. $512 = 8^2$
15. $96 \div 4 = 24$
    $24 \times 1 = 24$
16. $88 \div 8 = 11$
    $11 \times 3 = 33$
17. $100 \div 10 = 10$
    $10 \times 7 = 70$
18. $60 \div 4 = 15$
    $15 \times 3 = 45$ minutes
19. $24 \div 8 = 3$
    $3 \times 1 = 3$ apples
20. $20 \div 10 = 2$
    $2 \times 9 = 18$ people

Test 2
1. $10 \times 10 \times 10 \times 10 = 10,000$
2. $10 \times 10 = 100$
3. $10 \times 10 \times 10 = 1,000$
4. $10 = 10^1$
5. $1,000,000 = 10^6$
6. $1 = 10^0$
7. $1 \times 100 + 2 \times 10 + 5 \times 1$
8. $3 \times 1,000 + 9 \times 100 + 3 \times 10 + 2 \times 1$
9. $9 \times 10^3 + 2 \times 10^1 + 8 \times 10^0$
10. $1 \times 10^4 + 6 \times 10^3 + 9 \times 10^1$
11. 83,200
12. 2,680
13. $8 \times 8 = 64$
14. $2 \times 2 \times 2 = 8$
15. $5^3 = 125$
16. $\frac{1}{8} = \frac{2}{16} = \frac{3}{24} = \frac{4}{32}$
17. $\frac{4}{5} = \frac{8}{10} = \frac{12}{15} = \frac{16}{20}$
18. $24 \div 6 = 4$
    $4 \times 3 = 12$ hours
19. $24 - 12 = 12$ hours
20. 1/2 of 60 = 30 min
    1/3 of 60 = 20 min
    1/4 of 60 = 15 min
    1/4 of 60 = 15 min
    1/2 of 60 = 30 min
    $30 + 20 + 15 + 15 + 30 = 110$ min

Test 3
1. $3 \times 10^1 + 4 \times 10^0 + 6 \times \frac{1}{10^1}$
2. $1 \times \frac{1}{10^1} + 2 \times \frac{1}{10^2} + 4 \times \frac{1}{10^3}$
3. 5,417.04
4. 1,000.875
5. dimes; pennies;
   $3.00 + .0 + .05 = 3.05$
6. dollars; dimes; penny;
   $6.00 + .4 + .01 = 6.41$
7. $10 \times 10 \times 10 = 1,000$
8. $1 \times 1 \times 1 \times 1 \times 1 \times 1 \times 1 = 1$

TEST 3 - TEST 5

9. $6 \times 6 = 36$
10. $11 \times 11 = 121$
11. $\frac{2}{9} = \frac{4}{18} = \frac{6}{27} = \frac{8}{36}$
12. $\frac{1}{7} = \frac{2}{14} = \frac{3}{21} = \frac{4}{28}$
13. $\frac{3}{12} = \frac{1}{4}$
14. $\frac{2}{50} = \frac{1}{25}$
15. $\frac{8}{64} = \frac{1}{8}$
16. $\frac{15}{90} = \frac{1}{6}$
17. $4.63
18. $100 \div 10 = 10$
    $10 \times 5 = 50$ cents
19. $\frac{50}{100} = \frac{1}{2}$
20. $5 \times 5 = 5^2$ blocks

12. $\frac{5}{9} = \frac{10}{18} = \frac{15}{27} = \frac{20}{36}$
13. $\frac{1}{6} + \frac{2}{9} = \frac{9}{54} + \frac{12}{54} = \frac{21}{54} = \frac{7}{18}$
14. $\frac{1}{5} + \frac{7}{10} = \frac{10}{50} + \frac{35}{50} = \frac{45}{50} = \frac{9}{10}$
15. $\frac{1}{3} + \frac{3}{8} = \frac{8}{24} + \frac{9}{24} = \frac{17}{24}$
16. $4.75 + $6.29 = $11.04
    $11.04 > $11.00; yes
17. $1.2 + .75 + 1.15 = 3.1$ carats
18. $\frac{4}{10} = \frac{2}{5}$
19. $\frac{5}{8} + \frac{1}{6} = \frac{30}{48} + \frac{8}{48} = \frac{38}{48} = \frac{19}{24}$
20. $30 \div 5 = 6$
    $6 \times 3 = 18$ days

Test 4

1. $\overset{1}{6.7}$
    $\underline{+\ 5.4}$
    12.1

2. 2.0
    $\underline{+\ .2}$
    2.2

3. 6.24
    $\underline{+\ 8.40}$
    14.64

4. $\overset{1}{5.28}$
    $\underline{+2.05}$
    7.33

5. $1 \times 1 \times 1 = 1$
6. $10 \times 10 = 100$
7. $6 \times 6 \times 6 = 216$
8. $7 \times 7 = 49$
9. 8,400.2
10. 269.005
11. $\frac{4}{5} = \frac{8}{10} = \frac{12}{15} = \frac{16}{20}$

Test 5

1. $\overset{6}{\cancel{7}}.\overset{1}{2}5$
    $\underline{-\ 2.8}$
    4.45

2. $\overset{4}{\cancel{5}}.\overset{1}{0}7$
    $\underline{-\ 2.11}$
    2.96

3. $\overset{5}{\cancel{6}}.\overset{1}{\cancel{2}}\overset{1}{4}9$
    $\underline{-\ 5.371}$
    0.878

4. $\overset{1}{8.2}$
    $\underline{+\ .9}$
    9.1

5. $\overset{1}{6.14}$
    $\underline{+\ .395}$
    6.535

6. $\overset{1}{14.9}$
    $\underline{+\ .124}$
    15.024

7. $\frac{2}{5} + \frac{1}{8} = \frac{16}{40} + \frac{5}{40} = \frac{21}{40}$

8. $\frac{3}{7} + \frac{2}{9} = \frac{27}{63} + \frac{14}{63} = \frac{41}{63}$

9. $\frac{2}{3} + \frac{1}{6} = \frac{12}{18} + \frac{3}{18} = \frac{15}{18} = \frac{5}{6}$

10. 29,000.1

11. 580.034

12. $\frac{3}{4} - \frac{1}{5} = \frac{15}{20} - \frac{4}{20} = \frac{11}{20}$

13. $\frac{7}{8} - \frac{2}{3} = \frac{21}{24} - \frac{16}{24} = \frac{5}{24}$

14. $\frac{1}{4} - \frac{1}{6} = \frac{6}{24} - \frac{4}{24} = \frac{2}{24} = \frac{1}{12}$

15. $20.00 − $15.24 = $4.76

16. $7.35 + $2.35 + $1.17 + $1.00 = $11.87

17. $35.00 − $11.87 = $23.13

18. 3.2 + 2.3 + 1.05 = 6.55 hours

19. 100 ÷ 10 = 10
    10 × 2 = 20 cents

20. $\frac{3}{4} - \frac{1}{10} = \frac{30}{40} - \frac{4}{40} = \frac{26}{40} = \frac{13}{20}$

9. 25.32
   + 1.06
   ─────
   26.38

10. 7.³4¹2
    − 2. 3 9
    ───────
    5. 0 3

11. $\frac{1}{5} + \frac{1}{9} = \frac{9}{45} + \frac{5}{45} = \frac{14}{45}$

12. $\frac{3}{8} - \frac{1}{4} = \frac{12}{32} - \frac{8}{32} = \frac{4}{32} = \frac{1}{8}$

13. $\frac{7}{9} - \frac{2}{7} = \frac{49}{63} - \frac{18}{63} = \frac{31}{63}$

14. $5\frac{2}{3} = \frac{15}{3} + \frac{2}{3} = \frac{17}{3}$

15. $2\frac{6}{7} = \frac{14}{7} + \frac{6}{7} = \frac{20}{7}$

16. $1\frac{5}{8} = \frac{8}{8} + \frac{5}{8} = \frac{13}{8}$

17. kilometers

18. 1,000

19. 6.1 − 5.7 = .4 minutes

20. $3 \times 10^3 + 4 \times 10^2 + 1 \times \frac{1}{10^1} + 2 \times \frac{1}{10^2}$

Test 6

1. | kilogram(kg) | hectogram(hg) |
   |---|---|
   | 1,000 g | 100 g |
   | dekagram(dkg) | gram(g) |
   | 10 g | 1 g |

2. | kiloliter(kl) | hectoliter(hl) |
   |---|---|
   | 1,000 L | 100 L |
   | dekaliter(dkl) | liter(L) |
   | 10 L | 1 L |

3. | kilometer(km) | hectometer(hm) |
   |---|---|
   | 1,000 m | 100 m |
   | dekameter(dkm) | meter(m) |
   | 10 m | 1 m |

4. c
5. a
6. b
7. 1.2
   − .4
   ────
   .8

8. ⁴5.¹0
   − 2. 8
   ──────
   2. 2

Test 7

1. | gram(g) | decigram(dg) |
   |---|---|
   | 1 g | 1/10 g |
   | centigram(cg) | milligram(mg) |
   | 1/100 g | 1/1,000 g |

2. | liter(L) | deciliter(dl) |
   |---|---|
   | 1 liter | 1/10 liter |
   | centiliter(cl) | milliliter(ml) |
   | 1/100 liter | 1/1,000 liter |

3. | meter(m) | decimeter(dm) |
   |---|---|
   | 1 m | 1/10 m |
   | centimeter(cm) | millimeter(mm) |
   | 1/100 m | 1/1,000 m |

4. kilo
5. hecto
6. deka
7. 40 ÷ 2 = 20
   20 × 1 = 20
8. 64 ÷ 8 = 8
   8 × 3 = 24

9. $14 \div 7 = 2$
    $2 \times 2 = 4$
10. $\frac{17}{8} = \frac{16}{8} + \frac{1}{8} = 2\frac{1}{8}$
11. $\frac{35}{4} = \frac{32}{4} + \frac{3}{4} = 8\frac{3}{4}$
12. $\frac{29}{3} = \frac{27}{3} + \frac{2}{3} = 9\frac{2}{3}$
13. Pippin
14. 5 quarts
15. kilograms
16. kiloliter
17. half a mile
18. two pounds
19. $1.2 + 2.14 + 1.75 = 5.09$ in
20. $50.9 - 5.09 = 45.81$ in

Test 8

1. 45,000 liters
2. 6,000 mg
3. 1,300 mm
4. b
5. a
6. c
7. e
8. f
9. d
10. c
11. a
12. b
13. $6\frac{3}{4} + 3\frac{1}{5} = 6\frac{15}{20} + 3\frac{4}{20} = 9\frac{19}{20}$
14. $11\frac{1}{9} + 4\frac{2}{7} = 11\frac{7}{63} + 4\frac{18}{63} = 15\frac{25}{63}$
15. $21\frac{2}{5} + 7\frac{5}{6} = 21\frac{12}{30} + 7\frac{25}{30} =$
    $28\frac{37}{30} = 29\frac{7}{30}$
16. They are the same.
17. 1 kilogram
18. 5 grams
19. 500 centigrams
20. $4\frac{1}{2} + 4\frac{1}{2} = 8\frac{2}{2} = 9$ pies

Unit Test I

1. $1 \times 1 \times 1 \times 1 \times 1 \times 1 = 1$
2. $10 \times 10 \times 10 = 1,000$
3. $9 \times 9 = 81$
4. $3 \times 10 + 5 \times 1 + 2 \times \frac{1}{10} + 4 \times \frac{1}{100}$
5. $9 \times 10^4 + 1 \times \frac{10}{10^1} + 6 \times \frac{1}{10^2}$
6. 4,168.321
7. ${}^5 6.^1 3 9$
    $- 3.54$
    $ 2.85$
8. ${}^4 5.^9 0^1 0$
    $- 2.15$
    $ 2.85$
9. $ 1.^7 8^9 0^1 1$
    $- .999$
    $ .802$
10. $ 8.3$
    $+ .4$
    $ 8.7$
11. $ 1.23$
    $+ .147$
    $ 1.377$
12. $ 91.2$
    $+ .608$
    $ 91.808$
13. $96 \div 8 = 12$
    $12 \times 7 = 84$
14. $81 \div 9 = 9$
    $9 \times 5 = 45$
15. $\frac{6}{11} = \frac{12}{22} = \frac{18}{33} = \frac{24}{44}$
16. $\frac{9}{10} = \frac{18}{20} = \frac{27}{30} = \frac{36}{40}$
17. $\frac{1}{6} - \frac{1}{7} = \frac{7}{42} - \frac{6}{42} = \frac{1}{42}$
18. $\frac{3}{4} - \frac{2}{5} = \frac{15}{20} - \frac{8}{20} = \frac{7}{20}$
19. $\frac{7}{9} - \frac{4}{9} = \frac{3}{9} = \frac{1}{3}$
20. $1\frac{2}{3} + 4\frac{1}{6} = 1\frac{12}{18} + 4\frac{3}{18} = 5\frac{15}{18} = 5\frac{5}{6}$

21. $10\frac{3}{9} + 7\frac{1}{4} = 10\frac{12}{36} + 7\frac{9}{36} =$
    $17\frac{21}{36} = 17\frac{7}{12}$
22. $34\frac{5}{6} + 6\frac{1}{8} = 34\frac{40}{48} + 6\frac{6}{48} =$
    $40\frac{46}{48} = 40\frac{23}{24}$
23. 1,200 cm
24. 900,000 cg
25. 220 ml
26. gram
27. kilogram
28. meter
29. $52.5 + 15.06 = 67.56$ mi
30. $\$50.00 - \$16.95 = \$33.05$

10. liter
11. kilometer
12. kilogram
13. $12\frac{1}{4} - 3\frac{1}{5} = 12\frac{5}{20} - 3\frac{4}{20} = 9\frac{1}{20}$
14. $9 - 2\frac{7}{8} = 8\frac{8}{8} - 2\frac{7}{8} = 6\frac{1}{8}$
15. $21\frac{1}{3} - 6\frac{3}{4} = 21\frac{4}{12} - 6\frac{9}{12} =$
    $20\frac{16}{12} - 6\frac{9}{12} = 14\frac{7}{12}$
16. 10 grams
17. $4.2 \times .2 = .84$ mi
18. $3.3 + 2.75 + 4.09 = 10.14$ lb
19. $4 - 1\frac{5}{6} = 3\frac{6}{6} - 1\frac{5}{6} = 2\frac{1}{6}$ pies
20. $25.6 - 19.8 = 5.8$ mi

Test 9

1. $\begin{array}{r} 1.3 \\ \times\ .2 \\ \hline .26 \end{array}$
2. $\begin{array}{r} 2.3 \\ \times 1.2 \\ \hline 4\ 6 \\ 2\ 3\ \ \\ \hline 2.7\ 6 \end{array}$
3. $\begin{array}{r} .7 \\ \times\ .1 \\ \hline .07 \end{array}$
4. $\begin{array}{r} 1.1 \\ \times\ .6 \\ \hline .66 \end{array}$
5. $\begin{array}{r} 1.4 \\ \times 1.2 \\ \hline 2\ 8 \\ 1\ 4\ \ \\ \hline 1.6\ 8 \end{array}$
6. $\begin{array}{r} 2.0 \\ \times\ .4 \\ \hline .80 \end{array}$
7. 250 hg
8. 800 cm
9. 310 ml

Test 10

1. $\begin{array}{r} 6.24 \\ \times\ .4 \\ \hline 2.496 \end{array}$
2. $\begin{array}{r} 4.4 \\ \times 1.7 \\ \hline 3.08 \\ 4.4\ \ \\ \hline 7.48 \end{array}$
3. $\begin{array}{r} .67 \\ \times .12 \\ \hline .0134 \\ .067\ \ \\ \hline .0804 \end{array}$
4. $\begin{array}{r} 7.18 \\ \times\ .7 \\ \hline 5.026 \end{array}$
5. $\begin{array}{r} 3.3 \\ \times 1.5 \\ \hline 1.65 \\ 3.3\ \ \\ \hline 4.95 \end{array}$
6. $\begin{array}{r} .85 \\ \times .01 \\ \hline .0085 \end{array}$
7. $9 \times 100,000 = 900,000$ cg

8. $48 \times 10{,}000 = 48{,}000$ mm
9. $12.17 + 147.09 = 159.26$
10. $5.13 + .26 = 5.39$
11. $18.7 - 4.8 = 13.9$
12. $4\frac{3}{4} + 1\frac{1}{10} = 4\frac{30}{40} + 1\frac{4}{40} = 5\frac{34}{40} = 5\frac{17}{20}$
13. $13\frac{1}{2} - 4\frac{1}{8} = 13\frac{8}{16} - 4\frac{2}{16} = 9\frac{6}{16} = 9\frac{3}{8}$
14. $20 - 10\frac{1}{5} = 19\frac{5}{5} - 10\frac{1}{5} = 9\frac{4}{5}$
15. $\frac{5}{6} \times \frac{1}{4} = \frac{5}{24}$
16. $\frac{3}{4\!\!\!/_{\,8}} \times \frac{\cancel{2}}{3} = \frac{1}{4}$
17. $\frac{5}{3\!\!\!/_{\,6}} \times \frac{\cancel{2}}{3} = \frac{5}{9}$
18. $4 \times \$9.24 = \$36.96$
19. $642 \times .6 = 385.2$ miles
20. $\$59.50 \times 1.75 \approx \$104.13$

14. 9.30
   × .01
   ────
   9 3 0
   0 0 0
   ──────
   .0 9 3 0

15. 3.5
   × 4.5
   ────
   2
   1 5 5
   2
   1 2 0
   ──────
   15.7 5

16. 2.56
   × .81
   ────
   2 5 6
   6 0 8
   ──────
   2.07 36

17. $\frac{\overset{15}{\cancel{30}}}{7} \times \frac{3}{\cancel{2}} = \frac{45}{7} = 6\frac{3}{7}$

18. $\frac{5}{3} \times \frac{34}{\cancel{5}} = \frac{34}{3} = 11\frac{1}{3}$

19. $\$15.96 \times .15 = \$2.394$
    $\$15.96 + 2.39 = \$18.35$

20. $20 \times .60 = 12$ miles

Test 11

1. .35
2. .06
3. .19
4. .15
5. .58
6. .07
7. $\frac{2}{100} = \frac{1}{50}$
8. $\frac{23}{100}$
9. $\frac{44}{100} = \frac{11}{25}$
10. $1 \div 5 = .20 = 20\%$
11. $1 \div 2 = .50 = 50\%$
12. $1 \div 4 = .25 = 25\%$
13. $3 \div 4 = .75 = 75\%$

Test 12

1. $\frac{300}{100} + \frac{75}{100} = \frac{375}{100} = 375\% = 3.75$
2. $\frac{100}{100} + \frac{40}{100} = \frac{140}{100} = 140\% = 1.40$
3. $\frac{600}{100} + \frac{90}{100} = \frac{690}{100} = 690\% = 6.90$
4. $\frac{200}{100} + \frac{50}{100} = \frac{250}{100} = 250\% = 2.50$
5. $.35 = \frac{35}{100} = \frac{7}{20}$
6. $.11 = \frac{11}{100}$
7. yard
8. ounce
9. inch
10. inch
11. pounds
12. mile

TEST 12 - TEST 14

13. quarts
14. $\dfrac{3}{5} \div \dfrac{1}{3} = \dfrac{9}{15} \div \dfrac{5}{15} = \dfrac{9 \div 5}{1} = \dfrac{9}{5} = 1\dfrac{4}{5}$
15. $\dfrac{3}{4} \div \dfrac{7}{8} = \dfrac{24}{32} \div \dfrac{28}{32} = \dfrac{24 \div 28}{1} = \dfrac{24}{28} = \dfrac{6}{7}$
16. $\dfrac{2}{5} \div \dfrac{3}{8} = \dfrac{16}{40} \div \dfrac{15}{40} = \dfrac{16 \div 15}{1} = \dfrac{16}{15} = 1\dfrac{1}{15}$
17. $\$11.25 \times 1.10 = \$12.38$ (rounded)
18. $25 \times 4.00 = 100$ cards
19. $\dfrac{1}{2} \div \dfrac{1}{10} = \dfrac{10}{20} \div \dfrac{2}{20} = \dfrac{10 \div 2}{1} = 5$ guests
20. $\$75.00 \times .25 = \$18.75$

Test 13

1.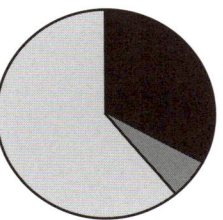
2. $120 \times .05 = 6$ minutes
3. cloudy
4. $30 \times .30 = 9$ days
5. $30 \times .40 = 12$ days
6. $\dfrac{400}{100} + \dfrac{75}{100} = \dfrac{475}{100} = 475\% = 4.75$
7. .25
8. .16
9. .06
10. 1.25
11. $3.61 \times .07 = .2527$
12. $.9 - .14 = .76$
13. $2.69 + 8.7 = 11.39$
14. $\dfrac{44}{5} \div \dfrac{14}{3} = \dfrac{132}{15} \div \dfrac{70}{15} = \dfrac{132 \div 70}{1}$

    $= \dfrac{132}{70} = \dfrac{66}{35} = 1\dfrac{31}{35}$
15. $\dfrac{17}{5} \div \dfrac{47}{7} = \dfrac{119}{35} \div \dfrac{235}{35} = \dfrac{119}{235}$
16. $\dfrac{29}{2} \div \dfrac{17}{3} = \dfrac{87}{6} \div \dfrac{34}{6} = \dfrac{87}{34} = 2\dfrac{19}{34}$
17. $\$1.45 \times 1.15 = \$1.67$ (rounded)
18. $10 \times 1{,}000 = 10{,}000$ grams
19. $1.5 \times 6 = 9$ books
20. A liter is close to 1 quart, so $4 \times 1 \approx 4$ liters.

Test 14

1. $(.1) \times (.5) = (.05)$

   ```
        .123
   ×    .456
   ------
        11
       628
      ¹115
       50
       12
      ¹48
   ------
     .056088
   ```

2. $(3) \times (6) = (18)$

   ```
         3.052
   ×     6.193
   -------
        ¹9 ¹156
       ¹2²7 4 1
          58
        3 0 5 2
       18 3 1 2
   -------
    18.9 0 1 0 3 6
   ```

3. $(.3) \times (.3) = (.09)$

   ```
         .281
   ×     .269
   ------
        729
       ²1 8
      ¹4 8 6
         2
        16 2
         4
   ------
    .075589
   ```

TEST 14 - TEST 16

4. $(2) \times (5) = (10)$

    $$\begin{array}{r} 1.504 \\ \times 5.167 \\ \hline {}^1\!3\,5\,2\,8 \\ 7 \\ {}^1\!3\,0\,2\,4 \\ 6 \\ {}^1\!1\,5\,0\,4 \\ 2\,5\,2\,0 \\ 5 \\ \hline 7.7\,7\,1\,1\,6\,8 \end{array}$$

5. 5
6. $\frac{1}{2}$
7. $\frac{38}{100} = \frac{19}{50}$
8. $\frac{2}{100} = \frac{1}{50}$
9. $\frac{900}{100} + \frac{40}{100} = \frac{940}{100} = 940\% = 9.40$
10. 16,000 mm
11. 350 hl
12. 10 dg
13. $\frac{5}{8} \div \frac{2}{3} = \frac{5}{8} \times \frac{3}{2} = \frac{15}{16}$
14. $\frac{1}{5} \div \frac{3}{8} = \frac{1}{5} \times \frac{8}{3} = \frac{8}{15}$
15. $\frac{1}{2} \div \frac{3}{5} = \frac{1}{2} \times \frac{5}{3} = \frac{5}{6}$
16. $10.4 \times \$1.639 = \$17.05$
17. $39.37 \times 3.2 = 125.984$ inches
18. $\$6.00 \times 1.20 = \$7.20$
19. $\$25.60 + \$11.19 + \$45.21 = \$82.00$
20. $\$82.00 \times 1.11 = \$91.02$

7. $(1) \times (.007) = (.007)$

    $$\begin{array}{r} 1.465 \\ \times .007 \\ \hline {}^1\!2\,4\,3 \\ 7\,8\,2\,5 \\ \hline 0.010\,2\,5\,5 \end{array}$$

8. $(200) \times (.1) = (20)$

    $$\begin{array}{r} 239.016 \\ \times .134 \\ \hline {}^1\!{}^1\!3 {}^2 \\ 8\,2\,{}^1\!6\,{}^1\!0\,4\,4 \\ {}^2 {}^1 \\ {}^3\!6\,9\,7\,0\,3\,8 \\ {}^1\!2\,3\,9\,0\,1\,6 \\ \hline 3\,2.0\,2\,8\,1\,4\,4 \end{array}$$

9. .06
10. .45
11. 1.30
12. 2.00
13. $34 \div 2 = 17$
14. $36 \div 9 = 4$
    $4 \times 2 = 8$
15. $100 \div 5 = 20$
    $20 \times 3 = 60$
16. $\frac{33}{8} \div \frac{6}{5} = \frac{\overset{11}{\cancel{33}}}{8} \times \frac{5}{\underset{2}{\cancel{6}}} = \frac{55}{16} = 3\frac{7}{16}$
17. $\frac{26}{3} \div \frac{1}{6} = \frac{26}{\cancel{3}} \times \frac{\overset{2}{\cancel{6}}}{1} = 52$
18. 1 kilogram
19. 500,000 mm
20. $35 \times 1.08 = 37.8$ tons

Test 15

1. .07 m
2. .1 kl
3. .5 g
4. .25 hm
5. 80 ml
6. 9,500 cg

Test 16

1. $3.14(10^2) = 314$ m^2
2. $3.14(20) = 62.8$ m
3. $3.14(2.5^2) = 19.625$ in^2
4. $3.14(5) = 15.7$ in
5. .061 m
6. .004 kg
7. 130 ml

TEST 16 - TEST 17

8. $.07 \times .04 = .0028$
9. $72.3 \times .9 = 65.07$
10. $204 \times .11 = 22.44$
11. $\dfrac{17}{2} \div \dfrac{9}{4} = \dfrac{17}{\cancel{2}_1} \times \dfrac{\cancel{4}^2}{9} = \dfrac{34}{9} = 3\dfrac{7}{9}$
12. $\dfrac{19}{6} \div \dfrac{5}{3} = \dfrac{19}{\cancel{6}_2} \times \dfrac{\cancel{3}}{5} = \dfrac{19}{10} = 1\dfrac{9}{10}$
13. $\dfrac{9}{10} \div \dfrac{2}{5} = \dfrac{9}{\cancel{10}_2} \times \dfrac{\cancel{5}}{2} = \dfrac{9}{4} = 2\dfrac{1}{4}$
14. $9 + 4 + 9 + 4 = 26$ m
15. $4.5 + 4.5 + 4.5 + 4.5 = 18$ ft
16. $6.2 + 3 + 6.2 + 3 = 18.4$ in
17. $3.14(6^2) = 113.04$ ft^2
18. $3.14(5) = 15.7$ ft
19. $8 + 8 + 8 + 8 = 32$ yd
20. $2 \times 100 = 200$ m

Unit Test II

1.
```
    8.6 1
  ×   .7
    1 4
  5 6 2 7
  6.0 2 7
```

2.
```
      1.4
    ×2.6
      2
    1 6 4
    2 8
    3.6 4
```

3.
```
      .18
    ×.95
    1 4
    7 5 0
    9 2
    .17 1 0
```

4.
```
        .106
      × .352
        2 1 2
      ¹5 3 0
      3 1 8
      .0 3 7 3 1 2
```

5.
```
        2.605
      × 6.279
        5
     ¹18 4 4 5
        4
     ¹14 2 3 5
        1
     ¹3 4 2 1 0
     1 2 6 3 0
     16.3 5 6 7 9 5
```

6.
```
        3.191
      × 4.260
        5 0
     ¹²8 6 4 6
        1
      6 2 8 2
        3
     1¹2 4 6 4
     13.5 9 3 6 6 0
```

7. .01
8. .10
9. 1.00
10. $\dfrac{5}{100} = \dfrac{1}{20}$
11. $\dfrac{48}{100} = \dfrac{12}{25}$
12. $\dfrac{16}{100} = \dfrac{4}{25}$
13. $.25 = 25\%$
14. $.40 = 40\%$
15. $.50 = 50\%$
16. $.75 = 75\%$
17. $\dfrac{600}{100} + \dfrac{25}{100} = \dfrac{625}{10} = 625\% = 6.25$
18. .2 m
19. 8,100,00 cg
20.

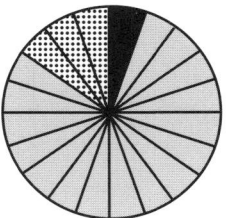

■ electricity

□ ingredients

▦ paper cups

UNIT TEST II - TEST 19

21. $\$35 \times .80 = \$28$
22. $3.14(3^2) = 28.26 \text{ m}^2$
23. $3.14(6) = 18.84 \text{ m}$
24. $\frac{27}{5} \times \frac{21}{5} = \frac{567}{25} = 22\frac{17}{25}$
25. $\frac{7}{3} \times \frac{5}{9} = \frac{35}{27} = 1\frac{8}{27}$
26. $\frac{15}{2} \div \frac{5}{4} = \frac{\cancel{15}^{3}}{\cancel{2}} \times \frac{\cancel{4}^{2}}{\cancel{5}} = 6$
27. $\frac{20}{3} \div \frac{2}{5} = \frac{\cancel{20}^{10}}{3} \times \frac{5}{\cancel{2}} = \frac{50}{3} = 16\frac{2}{3}$
28. $.88 \times 50 = 44$ questions
29. $\$135.00 \times 1.16 = \$156.60$
30. $10.16 + 9.25 + 10.16 + 9.25 = 38.82 \text{ in}$

Test 17

1. $2\overline{)\,.004} = .002$ (4, 0)
2. $.002 \times 2 = .004$
3. $8\overline{)\,1.76} = .22$ (16, 16, 0)
4. $.22 \times 8 = 1.76$
5. $4\overline{)\,.28} = .07$ (28, 0)
6. $.07 \times 4 = .28$
7. $5\overline{)\,1.0} = 0.2$ (10, 0)
8. $.2 \times 5 = 1.0$
9. $7\overline{)\,3.5} = 0.5$ (35, 0)
10. $.5 \times 7 = 3.5$
11. $3\overline{)\,.009} = .003$ (9, 0)
12. $.003 \times 3 = .009$
13. $3.14(1.2^2) = 4.5216 \text{ mi}^2$
14. $3.14(2.4) = 7.536 \text{ mi}$
15. $40 + 30 + 40 + 30 = 140 \text{ in}$
16. $7.9 + 3.2 + 7.9 + 3.2 = 22.2 \text{ m}$
17. $6\frac{1}{4} + 4\frac{3}{4} + 6\frac{1}{4} + 4\frac{3}{4} = 20\frac{8}{4} = 22 \text{ in}$
18. $\$40.95 \div 7 = \$5.85$
19. $71.6 \div 4 = 17.9 \text{ mi}$
20. $3.5 + 6.9 + 4.3 = 14.7 \text{ hr}$
    $14.7 \times \$7.10 = \$104.37$

Test 18

1. $.004\overline{)\,484.000} = 121,000$ (400,000; 84,000; 80,000; 4,000; 4,000; 0)

2. $\begin{array}{r} 121{,}000 \\ \times\ \ \ .004 \\ \hline 484.000 \end{array}$

3. $\begin{array}{r} 201 \\ .18\overline{)36.18} \\ \underline{3600} \\ 18 \\ \underline{18} \\ 0 \end{array}$

4. $\begin{array}{r} 2\ 0\ 1 \\ \times\ \ \ .18 \\ \hline 1\ 6\ 0\ 8 \\ 2\ 0\ 1 \\ \hline 36.18 \end{array}$

5. $\begin{array}{r} 3.7 \\ 6\overline{)22.2} \\ \underline{18.0} \\ 42 \\ \underline{42} \\ 0 \end{array}$

6. $\begin{array}{r} 3.7 \\ \times\ \ 6 \\ \hline 1\ 4 \\ 1\ 8\ 2 \\ \hline 22.2 \end{array}$

7. $\begin{array}{r} .0006 \\ 9\overline{).0054} \\ \underline{54} \\ 0 \end{array}$

8. $\begin{array}{r} .0006 \\ \times\ \ \ \ \ 9 \\ \hline .0054 \end{array}$

9. $3.14(3.5^2) = 38.465\ \text{in}^2$
10. $3.14(7) = 21.98\ \text{in}$
11. $13 \times 13 = 169$
12. $6 \times 6 = 36$
13. $10 \times 10 = 100$
14. $11 + 11 + 12 = 34\ \text{ft}$
15. $9 + 12 + 15 = 36\ \text{in}$
16. $6.9 + 5.4 + 4.1 = 16.4\ \text{m}$
17. $\$19 \div .25 = 76\ \text{quarters}$
18. $5.4 \div .45 = 12\ \text{batches}$
19. $13.9 + 13.9 + 13.9 + 13.9 = 55.6\ \text{m}$
20. $3{,}160 \div 1{,}000 = 3.16\ \text{kg}$

Test 19

1. $.19G = 38$
   $G = 38 \div .19$
   $G = 38 \div 200$
2. $.19(200) = 38$
   $38 = 38$
3. $.008Y = 2$
   $Y = 2 \div .008$
   $Y = 250$
4. $.008(250) = 2$
   $2 = 2$
5. $.6G = 24$
   $G = 24 \div .6$
   $G = 40$
6. $.6(40) = 24$
   $24 = 24$
7. $.11Y = 55$
   $Y = 55 \div .11$
   $Y = 500$
8. $.11(500) = 55$
   $55 = 55$
9. $\begin{array}{r} 1{,}003 \\ 006.\overline{)6\ 018.} \\ \underline{6{,}000} \\ 18 \\ \underline{18} \\ 0 \end{array}$

10. $\begin{array}{r} 1{,}003 \\ \times\ \ .006 \\ \hline 6.018 \end{array}$

11. $\begin{array}{r} 0.2 \\ 15\overline{)3.0} \\ \underline{30} \\ 0 \end{array}$

12. $\begin{array}{r} 15 \\ \times\ \ .2 \\ \hline 3.0 \end{array}$

13. $8 \times 4 = 32\ \text{m}^2$
14. $3.9 \times 3.9 = 15.21\ \text{ft}^2$
15. $5.1 \times 2.5 = 12.75\ \text{in}^2$
16. $.45J = 90$
   $J = 90 \div .45$
   $J = 200\ \text{hours}$

TEST 19 - TEST 20

17. 200 − 90 = 110 hours
18. 75G = 12
    G = 12 ÷ .75
    G = 16 gifts
19. $308.00 ÷ 8 = $38.50
20. 5,000 g = 5 kg
    50 kg > 5 kg

Test 20

1.
```
      55
  5.)275.
     250
      25
      25
       0
```

2.
```
    55
  ×  .5
    2
   255
  2.75
```

3.
```
        21.25
  32.)680.00
      640.00
       40.00
       32.00
        8.00
        6.40
        1.60
        1.60
           0
```

4.
```
    21.25
  ×   .32
    1 1 1
    4240
    1
    6365
    6.8000
```

5.
```
           680
   04.)2720.
       2400
        320
        320
          0
```

6.
```
     680
  ×  .04
    3
    2420
    27.20
```

7.
```
         15
   28.)420.
       280
       140
       140
         0
```

8.
```
     .28
  × 15
    1
    140
    28
    4.20
```

9.
```
         .056
   35)1.960
      1750
       210
       210
         0
```

10.
```
     .056
  ×   35
    13
    250
    1
    158
    1.960
```

11.
```
          4,270
   002.)8,540.
        8,000
          540
          400
          140
          140
            0
```

12. $\quad\quad 4\,2\,7\,0$
$\quad\quad\underline{\times\quad.002}$
$\quad\quad\quad\quad 1$
$\quad\quad\underline{8\,4\,4\,0}$
$\quad\quad 8.5\,4\,0$

13. $.12Q = .36$
$Q = .36 \div .12$
$Q = 3$
$.12(3) = .36; .36 = .36$

14. $2.5W = .75$
$W = .75 \div 2.5$
$W = .3$
$2.5(.3) = .75; .75 = .75$

15. $\frac{17}{3} \div \frac{6}{7} = \frac{17}{3} \times \frac{7}{6} = \frac{119}{18} = 6\frac{11}{18}$

16. $23\frac{10}{10} - 12\frac{1}{10} = 11\frac{9}{10}$

17. $\frac{4}{\cancel{5}} \times \frac{\cancel{10}^2}{11} = \frac{8}{11}$

18. $10 \times 6 = 60$ in^2

19. $27.3 \div 9.1 = 3$ parts

20. $\$66.36 \div \$3.16 = 21$ items

Test 21

1. $\quad\quad\quad 21.325 = 21.33$
$\quad 4.\overline{)8,5.300}$
$\quad\quad\quad \underline{8,0000}$
$\quad\quad\quad\quad 5300$
$\quad\quad\quad\quad \underline{4000}$
$\quad\quad\quad\quad 1300$
$\quad\quad\quad\quad \underline{1200}$
$\quad\quad\quad\quad\quad 100$
$\quad\quad\quad\quad\quad \underline{80}$
$\quad\quad\quad\quad\quad 20$

2. $\quad\quad\quad .245 = .25$
$\quad 16\overline{)3.930}$
$\quad\quad\quad \underline{3.200}$
$\quad\quad\quad\quad 730$
$\quad\quad\quad\quad \underline{640}$
$\quad\quad\quad\quad\quad 90$
$\quad\quad\quad\quad\quad \underline{80}$
$\quad\quad\quad\quad\quad 10$

3. $\quad\quad\quad 2.066 = 2.07$
$\quad 09.\overline{)18.600}$
$\quad\quad\quad \underline{18000}$
$\quad\quad\quad\quad 600$
$\quad\quad\quad\quad \underline{540}$
$\quad\quad\quad\quad\quad 60$
$\quad\quad\quad\quad\quad \underline{54}$
$\quad\quad\quad\quad\quad\quad 6$

4. $\quad\quad\quad 7.4545 = 7.\overline{45}$
$\quad 11.\overline{)82.0000}$
$\quad\quad\quad \underline{770000}$
$\quad\quad\quad\quad 50000$
$\quad\quad\quad\quad \underline{44000}$
$\quad\quad\quad\quad\quad 6000$
$\quad\quad\quad\quad\quad \underline{5500}$
$\quad\quad\quad\quad\quad\quad 500$
$\quad\quad\quad\quad\quad\quad \underline{440}$
$\quad\quad\quad\quad\quad\quad\quad 60$
$\quad\quad\quad\quad\quad\quad\quad \underline{55}$
$\quad\quad\quad\quad\quad\quad\quad\quad 5$

5. $\quad\quad\quad 44.0909 = 44.\overline{09}$
$\quad 22.\overline{)970.0000}$
$\quad\quad\quad \underline{8800000}$
$\quad\quad\quad\quad 900000$
$\quad\quad\quad\quad \underline{880000}$
$\quad\quad\quad\quad\quad 20000$
$\quad\quad\quad\quad\quad \underline{19800}$
$\quad\quad\quad\quad\quad\quad 200$
$\quad\quad\quad\quad\quad\quad \underline{198}$
$\quad\quad\quad\quad\quad\quad\quad 2$

6. $\quad\quad\quad 1.166 = 1.1\overline{6}$
$\quad 6\overline{)7.000}$
$\quad\quad \underline{6000}$
$\quad\quad\quad 1000$
$\quad\quad\quad \underline{600}$
$\quad\quad\quad\quad 400$
$\quad\quad\quad\quad \underline{360}$
$\quad\quad\quad\quad\quad 40$
$\quad\quad\quad\quad\quad \underline{36}$
$\quad\quad\quad\quad\quad\quad 4$

7. $30X = .8$
$X = .8 \div 30$
$X = .02\frac{2}{3}$

8. $.5Y = .123$
    $Y = .123 \div .5$
    $Y = .24\frac{3}{5}$
9. $2.3F = 6.41$
    $F = 6.41 \div 2.3$
    $F = 2.78\frac{16}{23}$
10. $65 - .45 = 64.55$
11. $7.03 + .2 = 7.23$
12. $200 - .02 = 199.98$
13. $1/2(12 \times 18) = 108$ ft^2
14. $1/2(5 \times 12) = 30$ in^2
15. $1/2(10.2 \times 4.1) = 20.91$ m^2
16. $51.84 \div 5.4 = 9.6$ m
17. $\$25.45 \times 1.20 = \$30.54$
18. $3.14(8^2) = 200.96$ in^2
19. grams
20. $31 \times 1{,}000 = 31{,}000$ g
    $31{,}000 - 310 = 30{,}690$ g

Test 22

1. $.09G + .38 = .425$
    $.09G = .425 - .38$
    $.09G = .045$
    $G = .045 \div .09 = .5$
2. $.09(.5) + .38 = .425$
    $.045 + .38 = .425$
    $.425 = .425$
3. $.4X + 3.6 = 6.4$
    $.4X = 6.4 - 3.6$
    $.4X = 2.8$
    $X = 2.8 \div .4 = 7$
4. $.4(7) + 3.6 = 6.4$
    $2.8 + 3.6 = 6.4$
    $6.4 = 6.4$
5. $1.5X + 4 = 4.165$
    $1.5X = 4.165 - 4$
    $1.5X = .165$
    $X = .165 \div 1.5 = .11$
6. $1.5(.11) + 4 = 4.165$
    $.165 + 4 = 4.165$
    $4.165 = 4.165$

7. $.944 = .94$
    $9 \overline{)8.500}$
    $\underline{8100}$
    400
    $\underline{360}$
    40
    $\underline{36}$
    4
8. $4.032 = 4.03$
    $4 \overline{)16.130}$
    $\underline{16000}$
    130
    $\underline{120}$
    10
    $\underline{8}$
    2
9. $8.971 = 8.97$
    $07 \overline{)62.800}$
    $\underline{56000}$
    6800
    $\underline{6300}$
    500
    $\underline{490}$
    10
    $\underline{7}$
    3
10. 1.8 m
11. $.549$ kl
12. 45 mg
13. $2.8 \times 2.8 \times 2.8 = 21.952$ ft^3
14. $\frac{5}{4} \times \frac{5}{4} \times \frac{5}{4} = \frac{125}{64} = 1\frac{61}{64}$ m^3
15. $3 \times 22 = 66$ in^2
16. $22 + 4 + 22 + 4 = 52$ in
17. $\frac{1}{2}(4 \times 6) = 12$ m^2
18. $5.4 + 4.2 + 6 = 15.6$ m
19. $\$1.69 \times 6 = \$10.14$
    $\$10.14 \times 1.03 = \$10.44$
20. $.45D + 1.35 = 23.85$
    $.45D = 23.85 - 1.35$
    $.45D = 22.50$
    $D = 22.50 \div .45 = \$50$

Test 23

1. $3 \div 10 = .30 = 30\%$
2. $1 \div 7 = .14\frac{2}{7} = 14\frac{2}{7}\%$
3. $2 \div 3 = .66\frac{2}{3} = 66\frac{2}{3}\%$
4. $5 \div 6 = .83\frac{1}{3} = 83\frac{1}{3}\%$
5. $1 \div 13 = .07\frac{9}{13} = 7\frac{9}{13}\%$
6. $8 \div 9 = .88\frac{8}{9} = 88\frac{8}{9}\%$
7. $.9R + .5 = .77$
   $.9R = .77 - .5$
   $.9R = .27$
   $R = .27 \div .9 = .3$
8. $.9(.3) + .5 = .77$
   $.27 + .5 = .77$
   $.77 = .77$
9. $5.2833 = 5.28\overline{3}$
   $6\overline{|31.7000}$
   $\underline{300000}$
   $\,17000$
   $\,\underline{12000}$
   5000
   $\underline{4800}$
   200
   $\underline{180}$
   2
10. $303.33 = 303.\overline{3}$
    $3\overline{|910.00}$
    $\underline{90000}$
    $\,1000$
    $\,\underline{900}$
    $\,\,100$
    $\,\,\underline{90}$
    $\,\,\,10$
    $\,\,\,\underline{9}$
    $\,\,\,\,1$
11. $4.722 = 4.7\overline{2}$
    $09\overline{|42.500}$
    $\underline{36000}$
    $\,6500$
    $\,\underline{6300}$
    $\,\,200$
    $\,\,\underline{180}$
    $\,\,\,20$
    $\,\,\,\underline{18}$
    $\,\,\,\,2$
12. $8 \times 9 \times 5 = 360 \text{ in}^3$
13. $3.4 \times 2.1 \times 1 = 7.14 \text{ m}^3$
14. $\frac{1}{2} \times \frac{1}{2} \times \frac{3}{4} = \frac{3}{16} \text{ ft}^3$
15. $\frac{35}{50} = \frac{70}{100} = 70\%$
16. $1 \div 8 = .12\frac{1}{2} = 12\frac{1}{2}\%$
17. $\$5.65 \times .60 = \$3.39$
18. $45 \times .80 = 36$ done
    $45 - 36 = 9$ toys left
19. $\$11.20 \times 1.20 = \$13.44$
    $\$20.00 - \$13.44 = \$6.56$
20. $100 \div 10 = 10$ ft

Unit Test III

1. $.002$
   $4\overline{|.008}$
   $\underline{8}$
   0
2. $.002$
   $\underline{\times 4}$
   $.008$
3. $.05$
   $6\overline{|.30}$
   $\underline{30}$
   0
4. $.05$
   $\underline{\times 6}$
   $.30$

5.
$$2.\overline{)860.} 430$$
   ```
        430
   2.|860.
       800
        60
        60
         0
   ```

6.
   ```
      430
   ×   .2
     86.0
   ```

7.
   ```
            .07
   09.|00.63
         63
          0
   ```

8.
   ```
      .07
   × .09
    .0063
   ```

9.
   ```
           16
   21.|336.
       210
       126
       126
         0
   ```

10.
   ```
      16
   ×2.1
    116
    22
   33.6
   ```

11.
   ```
            1,810
   005.|9,050.
        5,000
        4,050
        4,000
           50
           50
            0
   ```

12.
   ```
     1,810
   × .005
        4
      5 0 50
     9.050
   ```

13. $6.3W = 11.34$
    $W = 11.34 \div 6.3$
    $W = 1.8$

14. $6.3(1.8) = 11.34$
    $11.34 = 11.34$

15. $.7X + .08 = 17.58$
    $.7X = 17.58 - .08$
    $.7X = 17.5$
    $X = 17.5 \div .7 = 25$

16. $.7(25) + .08 = 17.58$
    $17.5 + .08 = 17.58$
    $17.58 = 17.58$

17. $2 \times 10 = 20 \text{ in}^2$

18. $10 + 5.5 + 10 + 5.5 = 31 \text{ in}$

19. $1.4 \times .7 = .98 \text{ m}^2$

20. $1.4 + .7 + 1.4 + .7 = 4.2 \text{ m}$

21. $\frac{1}{2}(3 \times 6) = 9 \text{ ft}^2$

22. $2 + 4 + 6 = 12 \text{ ft}$

23.
   ```
            12.233 = 12.23̄
      6|73.400
        60000
        13400
        12000
         1400
         1200
          200
          180
           20
   ```

24.
   ```
            15.166 = 15.16̄
      3.|45.500
        30000
        15500
        15000
          500
          300
          200
          180
           20
   ```

25. $8\overline{).700}.87\frac{4}{8}=.87\frac{1}{2}$
    $\phantom{8\overline{)}}\underline{640}$
    $\phantom{8\overline{)}}60$
    $\phantom{8\overline{)}}\underline{56}$
    $\phantom{8\overline{)}}4$

26. $9 \div 10 = .90 = 90\%$

27. $2 \div 7 = .28\frac{4}{7} = 28\frac{4}{7}\%$

28. $1 \div 3 = .33\frac{1}{3} = 33\frac{1}{3}\%$

29. $3 \div 4 = .75 = 75\%$

30. $8 \times 9 \times 7.2 = 518.4 \text{ ft}^3$

Test 24

1. $\frac{85}{100} = \frac{17}{20}$

2. $\frac{16}{100} = \frac{4}{25}$

3. $\frac{2}{10} = \frac{1}{5}$

4. $\frac{5}{1000} = \frac{1}{200}$

5. $2 \div 9 = .22\frac{2}{9} = 22\frac{2}{9}\%$

6. $4 \div 5 = .80 = 80\%$

7. $1 \div 2 = .50 = 50\%$

8. $2 \div 3 = .66\frac{2}{3} = 66\frac{2}{3}\%$

9. $.75X = 7.5$
    $X = 7.5 \div .75 = 10$

10. $.75(10) = 7.5$
    $7.5 = 7.5$

11. $1.4X + 5 = 6.82$
    $1.4X = 6.82 - 5$
    $1.4X = 1.82$
    $X = 1.82 \div 1.4 = 1.3$

12. $1.4(1.3) + 5 = 6.82$
    $1.82 + 5 = 6.82$
    $6.82 = 6.82$

13. $3\overline{)40.00}13.33\frac{1}{3}$
    $\phantom{3\overline{)}}\underline{3000}$
    $\phantom{3\overline{)}}1000$
    $\phantom{3\overline{)}}\underline{900}$
    $\phantom{3\overline{)}}100$
    $\phantom{3\overline{)}}\underline{90}$
    $\phantom{3\overline{)}}10$
    $\phantom{3\overline{)}}\underline{9}$
    $\phantom{3\overline{)}}1$

14. $11\overline{)26.80}2.43\frac{7}{11}$
    $\phantom{11\overline{)}}\underline{2200}$
    $\phantom{11\overline{)}}480$
    $\phantom{11\overline{)}}\underline{440}$
    $\phantom{11\overline{)}}40$
    $\phantom{11\overline{)}}\underline{33}$
    $\phantom{11\overline{)}}7$

15. $1 + 5 + 9 = 15$
    $15 \div 3 = 5$

16. $2 + 2 + 5 + 7 = 16$
    $16 \div 4 = 4$

17. $75 \times 1.30 = 97.5 \text{ lb}$

18. $45 + 31 + 45 + 31 = 152 \text{ in}$

19. $3.14(60)^2 = 11{,}304 \text{ ft}^2$

20. $3.5 \times 1{,}000 = 3{,}500 \text{ m}$

Test 25

1. $8 + 13 + 17 + 17 + 19 + 19 + 19 = 112$
    mean $= 112 \div 7 = 16$
    median $= 17$
    mode $= 19$

2. $70 + 80 + 80 + 90 + 100 = 420$
    mean $= 420 \div 5 = 84$
    median $= 80$
    mode $= 80$

3. $8 + 8 + 9 + 10 + 15 = 50$
    mean $= 50 \div 5 = 10$
    median $= 9$
    mode $= 8$

4. $3+3+6+7+11=30$
    mean $= 30 \div 5 = 6$
    median $= 6$
    mode $= 3$
5. $\dfrac{53}{100}$
6. $\dfrac{44}{100} = \dfrac{11}{25}$
7. $23 \div 40 = .57\dfrac{1}{2} = 57\dfrac{1}{2}\%$
8. $4 \div 6 = .66\dfrac{2}{3} = 66\dfrac{2}{3}\%$
9. $9X + 3.4 = 7.9$
    $9X = 7.9 - 3.4$
    $9X = 4.5$
    $X = 4.5 \div 9 = .5$
10. $9(.5) + 3.4 = 7.9$
    $4.5 + 3.4 = 7.9$
    $7.9 = 7.9$
11. $10 \times 10 = 100$
12. $3 \times 3 \times 3 = 27$
13. $1 \times 1 \times 1 \times 1 \times 1 \times 1 = 1$
14. $.9 + 1 + .9 + 1 = 3.8$ m
15. $.9 \times .8 = .72$ m^2
16. $9.7 + 9.7 + 9.7 + 9.7 = 38.8$ ft
17. $9.7 \times 9.7 = 94.09$ ft^2
18. $2+2+3+3+4+4+4+6+8 = 36$
    average, or mean $= 36 \div 9 = 4$
19. most often, or mode $= 4$
20. middle, or median $= 4$

Test 26

1. $\dfrac{10}{1000} = \dfrac{1}{100}$
2. $\dfrac{300}{1000} = \dfrac{3}{10}$
3. $\dfrac{700}{1000} = \dfrac{7}{10}$
4. $\dfrac{4}{10} = \dfrac{2}{5}$
5. $\dfrac{4}{10} = \dfrac{2}{5}$
6. $\dfrac{7}{10}$

7. $3+7+8+8+9 = 35$
    mean $= 35 \div 5 = 7$
    median $= 8$
    mode $= 8$
8. $9+9+12+13+17 = 60$
    mean $= 60 \div 5 = 12$
    median $= 12$
    mode $= 9$
9. $\dfrac{5}{10} = \dfrac{1}{2}$
10. $\dfrac{92}{100} = \dfrac{23}{25}$
11. $\dfrac{9}{100}$
12. $\dfrac{825}{1000} = \dfrac{33}{40}$
13. $1 \div 3 = .33\dfrac{1}{3} = 33\dfrac{1}{3}\%$
14. $7 \div 8 = .87\dfrac{1}{2} = 87\dfrac{1}{2}\%$
15. $3.14(6.4) = 20$ ft
16. $200 \times .16 = 32$ people
17. 500 meters
18. $(8 + 4 - 6 + 3) \times 2 = 18$

Test 27

1. endpoint
2. length; width
3. ray
4. line segment
5. length; width
6. c
7. d
8. e
9. b
10. a
11. $\overrightarrow{XY}$
12. $\overleftrightarrow{QR}$ or $\overleftrightarrow{RQ}$
13. $1+1+1+2+5 = 10$
    mean $= 10 \div 5 = 2$
14. median $= 1$
15. mode $= 1$
16. $2.05 \times 1.3 = 2.665$

17. $1.43 \div .11 = 13$
18. $1.75 + 2.25 = 4$
19. $\dfrac{1}{2}$
20. $\dfrac{50}{100} = \dfrac{1}{2}$
    $8 \div 2 = 4$
    $4 \times 1 = 4$ chores done

Test 28
1. point
2. plane
3. plane
4. equal
5. congruent
6. similar
7. ↔
8. ~
9. ∞
10. →
11. •
12. −
13. ▱
14. ≅
15. 50%
16. 25%
17. 75%
18. $33\dfrac{1}{3}\%$
19. 20%
20. $66\dfrac{2}{3}\%$

Test 29
1. rays
2. 1/4
3. 90
4. 360
5. vertex
6. similar

7. congruent
8. length; width
9. ∠RGX or ∠XGR
10. $\dfrac{5}{10} = \dfrac{1}{2}$
11. $\dfrac{8}{100} = \dfrac{2}{25}$
12. $\dfrac{565}{1000} = \dfrac{113}{200}$
13. $\dfrac{\cancel{2}4}{5} \times \dfrac{3}{\cancel{10}5} = \dfrac{6}{25}$
14. $\dfrac{3}{24} + \dfrac{14}{24} = \dfrac{17}{24}$
15. $\dfrac{5}{8} - \dfrac{4}{8} = \dfrac{1}{8}$
16. $\dfrac{5}{150} = \dfrac{1}{30}$
17. 9 is most often, or mode
18. $16.34 + 37.2 + 45.5 + 61 = 160.04$
    $160.04 \div 4 = 40.01$ lb

Test 30
1. 90°
2. acute
3. straight
4. obtuse
5. right
6. straight
7. obtuse
8. acute
9. 480%
10. 91%
11. 9%
12. $.25 \times 300 = 75$
13. $1.50 \times 80 = 120$
14. $.30 \times 75 = 22.5$
15. $\dfrac{2}{3} \div \dfrac{1}{8} = \dfrac{2}{3} \times \dfrac{8}{1} = \dfrac{16}{3} = 5\dfrac{1}{3}$
16. $\dfrac{4}{9} \div \dfrac{1}{3} = \dfrac{4}{\cancel{9}3} \times \dfrac{\cancel{3}}{1} = \dfrac{4}{3} = 1\dfrac{1}{3}$
17. $15 \times 15 = 225$
18. $3.14(12.2) = 36.308$ ft

Unit Test IV

1. $\dfrac{52}{100} = \dfrac{13}{25}$
2. $\dfrac{8}{100} = \dfrac{2}{25}$
3. $\dfrac{3}{10}$
4. $\dfrac{575}{1000} = \dfrac{23}{40}$
5. $10 + 10 + 11 + 12 + 17 = 60$
   mean $= 60 \div 5 = 12$
   median $= 11$
   mode $= 10$
6. $8 + 13 + 17 + 17 + 19 + 19 + 19 = 112$
   mean $= 112 \div 7 = 16$
   median $= 17$
   mode $= 19$
7. $\dfrac{40}{280} = \dfrac{1}{7}$
8. $\dfrac{210}{280} = \dfrac{3}{4}$
9. $\dfrac{40+30}{280} = \dfrac{70}{280} = \dfrac{1}{4}$
10. $\overline{FG}$ or $\overline{GF}$
11. $\overleftrightarrow{RS}$ or $\overleftrightarrow{SR}$
12. $\overrightarrow{AW}$
13. $\angle HJK$ or $\angle KJH$
14. ray
15. line segment
16. length; width
17. point
18. plane
19. congruent
20. equal
21. similar
22. vertex
23. 360
24. 90°
25. acute
26. straight
27. obtuse
28. c
29. e
30. f
31. b
32. d
33. g
34. a

Final Exam

1. $1 \times 1 \times 1 \times 1 \times 1 \times 1 = 1$
2. $8 \times 8 = 64$
3. $10 \times 10 \times 10 = 1000$
4. $5{,}271.349$
5. 7.452
   -1.85
   5.67
6. 6.0
   $+5.28$
   11.28
7. 32.041
   -0.596
   31.445
8. 2.49
   $\times.6$
   25
   1244
   1.494
9. 1.7
   $\times3$
   2
   31
   5.1
10. $.004$
    $\times.05$
    $.00020$
11. $1{,}300{,}000$ cm
12. $.25$ g
13. $.05$
14. $.65$
15. $\dfrac{25}{100} = \dfrac{1}{4}$
16. $\dfrac{32}{100} = \dfrac{8}{25}$
17. $.8 = 80\%$

18. $5 \div 6 = .83\frac{1}{3} = 83\frac{1}{3}\%$

19. $\frac{400}{100} + \frac{60}{100} = \frac{460}{100} = 4.60 = 460\%$

20. $\frac{78}{100} = \frac{39}{50}$

21. $\frac{3}{100}$

22.
```
        3.332  = 3.33
   4 ) 13.300
        1200
         130
         120
          10
           8
           2
```

23.
```
         .654  = .65
    7 ) 4.580
        4200
         380
         350
          30
          28
           2
```

24.
```
          65.66  = 65.6̄
    .6 ) 39.400
         36000
          3400
          3000
           400
           360
            40
            36
             4
```

25.
```
            .733  = .7̄3̄
    .03 ) .02200
           2100
            100
             90
             10
              9
              1
```

26.
```
          .81   9/11
   11 ) 9.00
        880
         20
         11
          9
```

27.
```
         .55   5/9
    9 ) 5.00
        450
         50
         45
          5
```

28. $3.2X + .07 = 4.55$
    $3.2X = 4.55 - .07$
    $3.2X = 4.48$
    $X = 4.48 \div 3.2$
    $X = 1.4$

29. $3.2(1.4) + .07 = 4.55$
    $4.48 + .07 = 4.55$
    $4.55 = 4.55$

30. plane
31. line segment
32. point
33. length; width
34. ray
35. obtuse
36. acute
37. similar
38. 360
39. 90°
40. straight
41. congruent
42. $A = 3.14(3)^2 = 28.26$ ft^2
    $C = 3.24(6) = 18.84$ ft
43. $\$3.50 + \$5 + \$5 + \$7 + \$8 = \$28.50$
    mean $= \$28.50 \div 5 = \$5.70$
    median $= \$5$
    mode $= \$5$
44. $\$45.60 \times 1.14 \approx \$51.98$
45. $\frac{1}{758}$

Word Problems 6-2

1. $3.45 + $1.99 + $6.59 + $12.98 = $25.01
   $25.01 - $2.50 = $22.51
   $50.00 - $22.51 = $27.49 is money left

2. Use a drawing to show their travels. The distances don't have to be to scale.

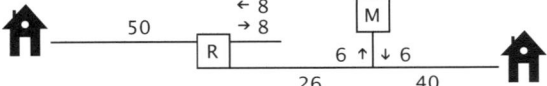

   50 + 26 + 40 = 116 miles from house to house
   50 + 8 + 8 = 66 miles past restaurant and back
   66 + 26 = 92 miles to museum turnoff
   92 + 6 + 6 = 104 miles back to main route
   104 + 40 = 144 miles total driven

3. 3.5 - .6 = 2.9"
   2.9 + 8.3 = 11.2"
   11.2 - 4.2 = 7" remaining

Word Problems 12-2

All money is rounded to hundredths at each step

1. 8 x $8.40 = $67.20
   7 x $5.99 = $41.93
   $67.20 + $41.93 = $109.13 discountable
   $109.13 x .10 = $10.91 (rounded)
   $109.13 - $10.91 = $98.22
   for discounted yarn
   2 x $2.50 = $5.00 for un-discounted
   $98.22 + $5.00 = $103.22 cost of yarn
   $103.22 x 1.13 = $116.64
   with tax and shipping (5% + 8% = 13%)

2. $5.99 x 5 = $29.95
   .10 x $29.95 = 2.995 rounds to $3.00
   $29.95 - $3.00 = $26.95 discounted price
   $26.95 x 1.13 = $30.45
   with tax and shipping
   You could also figure the total cost of one skein and multiply by 5. Rounding may give a slightly different answer.

3. $2.50 x 1.13 = 2.83
   for one skein (rounded)
   $2.83 x 1.5 = $4.25 (rounded)
   You could have found the basic cost of 1.5 skeins first, and then figured tax and shipping. Your answer may be slightly different because of rounding.

Word Problems 18-3

1. 1/4 + 1/2 + 3/4 + 1/4 =
   1/4 + 2/4 + 3/4 + 1/4 =
   7/4 or 1 3/4 pizza left over
   7/4 ÷ 1/4 = 7/4 x 4/1 = 7 boys

2. 5.3 ÷ 10 = .53"
   water from average snow
   4.1 ÷ 5 = .82" water from wet snow

   If you are unsure whether to multiply or divide, check your answer to see if it makes sense. The actual snowfall in each case was less than what was needed to make an inch of water, so division yields a sensible answer.
   .53 + .82 + 1.5 = 2.85"
   rounds to 2.9" of water

3. $(2)3.14 \times 3 = 18.84'$
    circumference of small garden
    $2 \times 3 = 6'$ diameter of small garden
    $3.14 \times 12 = 37.68'$ circumference
    of garden with doubled diameter
    $37.68 - 18.84 = 18.84'$
    additional edging needed

    In real life this would probably be rounded to 19 or 20 feet.

4. $3.14 \times 3^2 = 28.26$ ft^2 area of small garden
    28 ft^2 rounded
    $3.14 \times 6^2 = 113.04$ ft^2 area of large garden
    113 ft^2 rounded
    $113 \div 28 = 4.04$ or
    4 seed packets (rounded)
    When you have learned how to divide a decimal by a decimal, try this again using the unrounded areas.

Word Problems 24-2

1. $8 \times 12 = 96$ in^2 area of paper
    $3.14(3)^2 = 28.26^2$ in area of one circle
    $3.14(2.5)^2 = 19.625$ or 19.63 in^2
    rounded area of other circle
    $28.26 + 19.63 = 47.89$ in^2
    used for circles
    $96 - 47.89 = 48.11$ in^2 left over
    Look at the drawing to see why it is not possible to cut another circle with a 3" radius, even though there seems to be enough area.
    One circle has a diameter of 6", which leaves 2" distance to the edge of the paper.

    The other circle has a diameter of 5", which leaves a 3" distance to the edge of the paper.
    Neither space is enough for another circle with a 3" radius (6" diameter).

    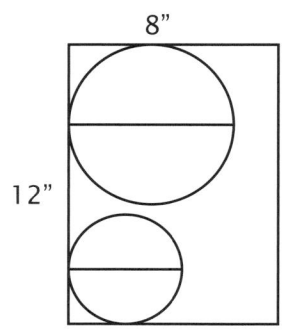

2. $\$6.50 \times .10 = \$.65$ is 10% of his hourly income
    $\$6.50 - .65 = \$5.85$
    hourly amount available to spend
    $\$25.35 + \$50.69 + \$85.96 =$
    $\$162$ total needed
    $\$162 \div \$5.85 = 27.69...$
    rounds to 28 hours

3. $.3 \times T = 300$ blue toys
    $T = 300 \div .3$
    $T = 1000$ toys in all
    $1/4 \times 1000 = 250$ green toys
    $300 + 250 = 550$ blue or green toys
    $1000 - 550 = 450$ remaining toys
    $1/2 \times 450 = 225$ red toys
    $1/2 \times 450 = 225$ yellow toys
    You could use decimals or fractions for this problem.

Symbols & Tables

IMPERIAL MEASUREMENT

3 teaspoons (tsp) = 1 tablespoon (Tbsp)
2 pints (pt) = 1 quart (qt)
8 pints = 1 gallon (gal)
4 quarts = 1 gallon
12 inches (in) = 1 foot (ft)
3 feet = 1 yard (yd)
5,280 feet = 1 mile (mi)
16 ounces (oz) = 1 pound (lb)
2,000 pounds = 1 ton

METRIC MEASUREMENT

1,000 millimeters (mm) - 1 meter (m)
100 centimeters (cm) - 1 meter
10 decimeters (dm) - 1 meter
10 meters = 1 dekameter (dkm or dam)
100 meters = 1 hectometer (hm)
1,000 meters = 1 kilometer (km)

Replace meter with liter or gram for liquid and weight equivalents.

METRIC – IMPERIAL EQUIVALENTS

1 meter (39.37 inches)
1 yard (36 inches)
1 centimeter ≈ .4 inch
1 liter ≈ 1.06 quarts
1 kilogram ≈ 2.2 pounds
1 kilometer ≈ .6 miles
1 inch ≈ 2.5 centimeters
1 ounce ≈ 28 grams

SYMBOLS

= equals
~ similar
≅ congruent
≈ approximately equal to
< less than
> greater than
% percent
π pi (22/7 or 3.14)
r^2 r squared or r · r
' foot
" inch
∞ infinite or infinity
· point
↔ line
→ ray
∠ angle
▱ plane
⌐ right angle

PERIMETER

rectangle, square, triangle, parallelogram - add the lengths of all the sides

CIRCUMFERENCE

circle $C = 2\pi r$

ZETA

AREA
rectangle, square, parallelogram
$A = bh$ (base times height)

triangle $\quad A = \dfrac{bh}{2}$

circle $\quad A = \pi r^2$

VOLUME
rectangular solid, cube
$V = Bh$ (area of base times height)

Glossary

A

Acute angle - an angle with a measure greater than 0° and less than 90°

Angle - a geometric figure formed by two rays joined at their origins

Area - the number of square units in a rectangle or other two-dimensional figure

Average - the result of adding a series of numbers and dividing by the number of items in the series

B–C

Base - the top or bottom side of a shape

Base 10 - the number system commonly used in this country

Centi - in the metric system, the Latin prefix representing one-hundredth

Circle - formed by a line drawn at an equal distance from a center point

Circumference - the distance around a circle. It corresponds to perimeter.

Coefficient - a number directly in front of an unknown. It is multiplied by the unknown.

Congruent - having the same shape and size

Cube - a three-dimensional figure with each side the same length

D

Deci - in the metric system, the Latin prefix representing one-tenth

Decimal or decimal fraction - a fraction written on one line by using a decimal point and place value

Decimal point - the dot used in decimal numbers between the units place and the tenths place

Decimal system - a number system based on 10, also called base 10

Deka or deca - in the metric system, the Greek prefix representing 10

Denominator - the bottom number in a fraction. It tells how many total parts there are in the whole.

Diameter - a line across a circle that passes through the center point. It is twice the length of the radius.

Dividend - the number being divided in a division problem

Divisor - the number that is being divided by in a division problem

E

Endpoint - one of the starting or stopping points of a line segment

Equal - having the same numerical value

Equation - a number sentence in which the value of one side is equal to the value of the other side

Equivalent - having the same value

Estimation - used to get an approximate value of an answer

Even number - a number that ends in 0, 2, 4, 6, or 8. Even numbers are multiples of two.

Expanded notation - a way of writing numbers in which each amount is multiplied by its place value

Exponent - a small number written to the upper right of another number. It tells how many times the first number is to be multiplied times itself.

Exponential notation - a form of notation in which each place value is indicated by 10 with an exponent

F–G

Factors - the two sides of a rectangle or the numbers multiplied in a multiplication problem

Fraction - one number written over another to show part of a whole. A fraction can also indicate division.

Geometry - the measure of the earth. Plane geometry studies one- and two-dimensional figures. Solid geometry studies three-dimensional figures.

Gram - the basic unit of weight or mass in the metric system

H–I

Hecto - in the metric system, the Greek prefix representing one hundred

Height - the length of a line from the top to the bottom of a shape that forms a right angle with the base

Improper fraction - a fraction with a numerator larger than its denominator

Inverse - opposite. Multiplication and division are inverses, and addition and subtraction are inverses.

Infinite - continuing in the same way or direction forever

K–L

Kilo - in the metric system, the Greek prefix representing one thousand

Line - in geometry, an infinite number of connected points

Line segment - a measurable piece of a line. It has two endpoints.

Liter or litre - the basic unit of liquid measure in the metric system

M

Mean - the average of a group of numbers

Median - the middle number in a list of numbers, when they are arranged from smallest to largest

Meter or metre - the basic unit of linear measure in the metric system

Metric system - a system of measurement based on 10, used in most countries other than the United States

Milli - in the metric system, the Latin prefix representing one-thousandth

Mixed number - a number made up of a whole number and a fraction

Mode - the number that occurs most often in a list

N–O

Numerator - the top number in a fraction. It tells how many of the parts of a whole have been chosen.

Obtuse angle - an angle with a measure greater than 90° and less than 180°

Origin - the starting point of a ray

P–Q

Parallel - two lines in the same plane that never touch each other

Parallelogram - a shape with four sides formed by two sets of parallel lines.

Per cent - "per hundredth" - a way of writing fractions that have a denominator of 100. The symbol is %.

Perpendicular - refers to two lines that form a right angle where they meet

Pi - the Greek letter π, that has a value of approximately 22/7 or 3.14

Pie graph - a circular graph used to show what percent of different parts are in a whole

Place value - the position of a number that tells what value it is assigned

Place-value notation - a way of writing numbers used to emphasize the place value of each part

Plane - a flat, two-dimensional figure that extends infinitely in all directions

Point - the smallest unit of measure in geometry. It has no length or width.

Power - another name for an exponent. 2^3 can be read "two to the power of three."

Probability - the chance of getting a desired outcome, given all possible outcomes

Product - the answer to a multiplication problem

Proper fraction - a fraction with a numerator smaller than its denominator

Quotient - the answer to a division problem

R

Radius - the distance from the center of a circle to its edge

Ratio - the relationship between two numbers written as a fraction

Rational number - a number that can be expressed as a fraction or ratio

Ray - a geometric figure that starts at a definite point and extends infinitely in one direction

Reciprocal - a number in which the numerator and denominator have switched places. A number times its reciprocal equals one.

Rectangle - a shape with four "square corners" or right angles

Rectangular solid - a three-dimensional shape with each side or face shaped like a rectangle

Reducing - dividing the numerator and denominator of a fraction by the same number. The resulting fraction is equivalent but in "lower terms."

Right angle - 90° angle (square corner)

Rounding - writing a number as its closest ten, hundred, etc. in order to estimate

Rule of four - a method for finding the same, or common, denominator of two fractions

S

Similar - having the same shape but not the same size

Square - a rectangle with all four sides the same length

Straight angle - an angle with a measure of 180°

T–U

Triangle - a shape with three sides

Unknown - a number whose value we do not know. It is usually represented by some letter of the alphabet.

V–Z

Variable - the unknown in an expression or equation, usually represented by a letter

Vertex - the origin of the rays of an angle or the point of the angle

Volume - the number of cubic units in a three-dimensional shape

Master Index for General Math

This index lists the levels at which main topics are presented in the instruction manuals for *Primer* through *Zeta*. For more detail, see the description of each level at www.mathusee.com. (Many of these topics are also reviewed in subsequent student books.)

Addition
    facts .... Primer, Alpha
    multiple digit .... Beta
Additive inverse .... Epsilon
Angles .... Zeta
Area
    circle .... Epsilon, Zeta
    parallelogram .... Delta
    rectangle .... Primer, Gamma, Delta
    square .... Primer, Gamma, Delta
    trapezoid .... Delta
    triangle .... Delta
Average .... Delta
Circle
    area .... Epsilon, Zeta
    circumference .... Epsilon, Zeta
    recognition .... Primer, Alpha
Circumference .... Epsilon, Zeta
Common factors .... Epsilon
Composite numbers .... Gamma
Congruent .... Zeta
Counting .... Primer, Alpha
Decimals
    add and subtract .... Zeta
    change to percent .... Epsilon, Zeta
    divide .... Zeta
    from a fraction .... Epsilon, Zeta
    multiply .... Zeta
Division
    facts .... Delta multiple
    digit .... Delta
Estimation .... Beta, Gamma, Delta
Expanded notation .... Delta, Zeta
Exponential notation .... Zeta
Exponents .... Zeta
Factors .... Gamma, Delta, Epsilon
Fractions
    add and subtract .... Epsilon
    compare .... Epsilon
    divide .... Epsilon
    equivalent .... Gamma, Epsilon
    fractional remainders .... Delta, Epsilon
    mixed numbers .... Epsilon
    multiply .... Epsilon
    of a number .... Delta, Epsilon
    of one .... Delta, Epsilon
    rational numbers .... Zeta
    reduce .... Epsilon
    to decimals .... Epsilon, Zeta
    to percents .... Epsilon, Zeta
Geometry
    angles .... Zeta
    area ...... Primer, Gamma, Delta, Epsilon,
    .... Zeta
    circumference .... Epsilon, Zeta
    perimeter .... Beta
    plane .... Zeta
    points, lines, rays .... Zeta
    shape recognition .... Primer, Alpha
Graphs
    bar and line .... Beta
    pie .... Zeta
Inequalities .... Beta
Linear measure .... Beta, Epsilon
Mean, median, mode .... Zeta
Measurement equivalents .... Beta, Gamma
Metric measurement .... Zeta
Mixed numbers .... Epsilon
Money .... Beta, Gamma
Multiplication
    facts .... Gamma
    multiple digit .... Gamma
Multiplicative inverse .... Epsilon
Number recognition .... Primer
Number writing .... Primer
Ordinal numbers .... Beta
Parallel lines .... Delta
Parallelogram .... Delta
Percent
    decimal to percent .... Epsilon, Zeta
    fraction to percent .... Epsilon, Zeta
    greater than 100 .... Zeta

of a number .... Zeta
Perimeter .... Beta
Perpendicular lines .... Delta
Pi .... Epsilon, Zeta
Pie graph .... Zeta
Place value ....
    Primer, Alpha, Beta, Gamma, Delta, Zeta
Prime factorization .... Epsilon
Prime numbers .... Gamma
Probability .... Zeta
Rational numbers .... Zeta
Reciprocal .... Epsilon
Rectangle
    area .... Primer, Gamma, Delta
    perimeter .... Beta
    recognition .... Primer, Alpha
Rectangular solid .... Delta
Regrouping .... Beta, Gamma
Roman numerals .... Delta
Rounding .... Beta, Gamma, Delta
Sequencing .... Beta
Skip counting
    2, 5, 10 .... Primer, Alpha, Beta
    all .... Gamma
Solve for unknown ....
    ..Primer, Alpha, Gamma, Delta, Epsilon,
    .... Zeta
Square
    area .... Primer, Gamma, Delta
    perimeter .... Beta
    recognition .... Primer, Alpha
Subtraction
    facts .... Alpha
    introduction .... Primer, Alpha
    multiple digit .... Beta
Tally marks .... Primer, Beta
Telling time .... Primer, Alpha, Beta
Thermometers and Gauges .... Beta
Trapezoid .... Delta
Triangle
    area .... Delta
    perimeter .... Beta
    recognition .... Primer, Alpha
Volume of rectangular solid .... Delta

Zeta Index

Acute angle.. 30
Algebra-decimal inserts....................... 4, 9
Angles 29, 30
Area
    circle 16
    rectangle and square .... Student 19D
    parallelogram .... Student 20D
    triangle................................ Student 21D
Average .... Student 24D, 25
Base................ Student 19D, 20D, 21D
Base ten... 3
Circle 16
Circumference 16
Coefficient 19, 22
Common denominator .... Student 4D
Congruent 28
Cube............................... 1, Student 22D
Decimal point 3, 4, 10, 14, 17, 18, 20
Decimals
    add... 4
    change to percent....................... 11, 23
    defined 3
    divide 17, 18, 20, 21
    expanded notation 3
    inserts 4, 9
    multiply 9, 10, 14
    rational numbers 24
    subtract... 5
Diameter.. 16
Dividend 17
Divisor... 17
Endpoint.. 27
Equal 28
Equations, solving............................ 19, 22
Expanded notation 2, 3
Exponential notation 2, 3
Exponents.. 1, 2
Factor 9
Fractions
    add............................... Student 4D, 8D
    common denominator .... Student 4D
    divide .... Student 12D, 13D, 14D, 15D
    equivalent............................ Student 2D
    fractional remainders...................... 21
    improper .... Student 6D, 7D, 13D
    mixed numbers ....
    Student 6D, 7D, 8D, 9D, 11D, 13D, 15D
    multiply .... Student 10D, 11D
    of a number........................ Student 1D
    rational numbers 24
    reduce................................ Student 3D
    rule of four .... Student 4D, 12D
    subtract...................... Student 5D, 9D
    to decimals 23
    to percents 11, 23
Geometry............................. 27, 28, 29, 30
Height .... Student 19D, 20D, 21D
Imperial equivalents................................ 7
Infinite... 27
Interstate Highway System ... Student 27D, E
Inverse... 19
Invert and multiply .... Student 14D
Line 27
Line segment 27
Mean 25
Median 25
Mental Math .... Student 26D
Metric measurement
    conversions 8, 15
    gram 6
    liter 6
    meter 6
    origin 6
    prefixes 6, 7
Mixed numbers ....
    Student 6D, 7D, 8D, 9D, 11D, 13D, 15D
Mode 25
Money 3, 18, 20
Obtuse angle 30
Origin 27
Percent
    decimal to percent 11, 23
    fraction to percent 11, 23
    greater than 100............................. 12
    of a number 11
    pie graphs 13
Perimeter
    parallelogram .... Student 17D
    rectangle, square .... Student 16D
    triangle................................ Student 18D
Pi 16
Pie graph 13
Place value 2
Plane 28

Point 27
Power of a number 1
Probability 26
Product 9
Quotient 17
Radius 16
Ratio 24
Rational numbers 24
Ray 27
Reciprocal .... Student 14D
Rectangle .... Student 16D, 19D
Rectangular solid .... Student 23D
Remainders 21
Rounding .... Student 11C – #15
Right angle 29
Taxes 11, 12
Tips 11, 12
Similar 28
Solving for unknown 19, 22
Square .... 1, Student 16D, 19D
Straight angle 30
Vertex 29
Variable 22
Volume
    cube .... Student 22D
    rectangular solid .... Student 23D
Word problems
    multistep 6, 12, 18, 24
    tips 1

224

0509
ZETA